Data-driven BIM for Energy Efficient Building Design

This research book aims to conceptualise the scale and spectrum of Building Information Modelling (BIM) and artificial intelligence (AI) approaches in energy efficient building design and to develop its functional solutions with a focus on four crucial aspects of building envelop, building layout, occupant behaviour and heating, ventilation and air-conditioning (HVAC) systems. Drawn from theoretical development on the sustainability, informatics and optimisation paradigms in built environment, the energy efficient building design will be marked through the power of data and BIM-intelligent agents during the design phase. It will be further developed via smart derivatives to reach harmony in the systematic integration of energy efficient building design solutions, a gap that is missed in the extant literature and that this book aims to fill. This approach will inform a vision for the future and provide a framework to shape and respond to our built environment and how it transforms the way we design and build. By considering the balance of BIM, AI and energy efficient outcomes, the future development of buildings will be regenerated in a direction that is sustainable in the long run. This book is essential reading for those in the AEC industry as well as computer scientists.

Dr Saeed Banihashemi is Associate Professor and Postgraduate Program Director of Building and Construction Information Management in the School of Design and Built Environment, Faculty of Arts and Design, University of Canberra (UC), Australia.

Dr Hamed Golizadeh is Assistant Professor of Building and Construction Management at the University of Canberra, Australia.

Professor Farzad Pour Rahimian is Professor of Digital Engineering and Manufacturing at Teesside University, UK.

Spon Research

Publishes a stream of advanced books for built environment researchers and professionals from one of the world's leading publishers. The ISSN for the Spon Research programme is ISSN 1940–7653 and the ISSN for the Spon Research E-book programme is ISSN 1940–8005

For more information about this series, please visit: www.routledge.com

Data-driven BIM for Energy Efficient Building Design

Saeed Banihashemi, Hamed Golizadeh and Farzad Pour Rahimian

Routledge
Taylor & Francis Group
LONDON AND NEW YORK

First published 2023
by Routledge
4 Park Square, Milton Park, Abingdon, Oxon OX14 4RN

and by Routledge
605 Third Avenue, New York, NY 10158

Routledge is an imprint of the Taylor & Francis Group, an informa business

© 2023 Saeed Banihashemi, Hamed Golizadeh and Farzad Pour Rahimian

British Library Cataloguing-in-Publication Data
A catalogue record for this book is available from the British Library

ISBN: 978-1-032-07348-4 (hbk)
ISBN: 978-1-032-07554-9 (pbk)
ISBN: 978-1-003-20765-8 (ebk)

DOI: 10.1201/9781003207658

Typeset in Times New Roman
by Apex CoVantage, LLC

Saeed dedicates this book to Shiva, his lovely wife, and his parents, for all their devotion and care.

Contents

Figures

Tables

Abbreviations

2D	CAD 2-Dimensional Computer-Aided Drawing
3D	3-Dimensional
AEC	Architecture, Engineering and Construction
AI	Artificial Intelligence
API	Application Program Interface
ANN	Artificial Neural Network
ASCE	American Society of Civil Engineering
BIM	Building Information Modelling
CDA	Conditional Demand Analysis
CDE	Common Data Environment
CFD	Computational Fluid Dynamic
DBLink	Database Link
DT	Decision Tree
EED	Energy Efficient Design
EU	European Union
GA	Genetic Algorithm
GB	Green Building
GBS	Green Building Studio
GHG	Greenhouse Gas
HVAC	Heating, Ventilation and Air Conditioning
ICT	Information and Communication Technology
IES	Integrated Environmental Solution
IFC	Industry Foundation Class
IFP	Implication for Practice
IT	Information Technology
IS	Information System
LEED	Leadership in Energy and Environmental Design
LoD	Level of Development
MLP	Multilayer Perceptron
MSE	Mean Square Error
ODBC	Open Database Connectivity
PSO	Particle Swarm Optimisation
RC	Reinforced Concrete
SVM	Support Vector Machine
TMY	Typical Metrological Year

Preface

The progress and development of technology and of equipping the design process with digital tools and algorithms have immersed our building design in a huge amount of data. This provides a rich flow of information about architecture and construction as well as proposes new building design solutions in the BIM and AI era. We therefore propose this initiative to drive the sustainable design of building projects through presenting data-driven and BIM-integrated energy efficient design solutions; this imperative is missed in the literature, and this book aims to fill this void.

Using BIM can expedite the energy efficient design (EED) process and provides the opportunity for testing and assessing different design alternatives and materials selection that may impact on the energy performance of buildings. However, the lack of intelligent decision-making platforms, ideal interoperability and inbuilt practices of optimisation methods in BIM hinder the full diffusion of BIM into EED. This premise triggers a new research direction, known as the integration of AI into BIM-EED. AI can develop and optimise EED in an integrated BIM platform to represent an alternative solution for building design. However, very little is known about achieving this.

Hence, this book primarily intends to present a research-led approach in identifying the massive opportunities that BIM and AI offer to sustainable design and strategies. An energy efficient building can be conceptualised in its balanced view of building envelop, building layout, occupant behaviour and HVAC systems. This book therefore theorises what a data-driven approach entails for the EED of these components, how BIM and AI can be integrated in the planning and design process and what types of design approaches contribute to this process.

It is of particular intent to spark new ideas on BIM, EED and AI and the relevant gaps, potential and challenges and to provide a research monograph on how a theoretical framework can be developed to facilitate energy efficiency at an early design stage. This framework functions via developing AI-based active BIM in order to obtain initial estimates and perceptions of the energy consumption of buildings and to optimise the values through recommending smart changes in design typologies, elements and variables. Such an integrative approach leads to achieving a comprehensive and informed recognition of the detailed analysis of buildings in terms of their components and function. On top of this, smart assessment, appraisal, monitoring and control of semantic, spatial and typological hierarchies of buildings are coupled with strategic planning and design solutions in shifting from unsustainable paradigms in building design to more sustainable practices through data-driven BIM.

We undertake an exploratory approach in addressing this book in order to discover how a data-centric BIM approach can be steered toward EED. This is an inexplicable phenomenon that has not yet been examined clearly and comprehensively, and so we try to explain how BIM and AI will advise and enhance the planning and design outcomes. This will commence with extensive research to unfold the area and concepts and to develop the theoretical foundations of BIM and AI-integrated building design. It will further conceptualise its characteristics and elements, introduce resources and technologies for its collective implementation and analyse how to apply these in various building components. An in-depth case study method will then be conducted to bring the ideas into a research-led practical context. This will allow for a greater explanation of the presented framework and detailed analysis of BIM and AI technologies and processes through the real-life experience of the actors involved in professional residential building cases.

This book will be a topical and prominent contribution in the built environment context as it addresses the validity of insights into BIM and AI in energy efficient buildings. Enhancing BIM applicability in terms of EED optimisation, shifting the current practice of post-design energy analysis, mitigating less-integrated building design platforms and lowering levels of interoperability are among the main significant outcomes of this research monograph. Ultimately, this book heads toward higher diffusion levels of BIM and AI into EED, which contributes significantly to the current body of knowledge and its research and development effects on the industry.

1 Classics of Data-Driven BIM for Energy Efficient Design

1.1 Background

Buildings have an enormous and continuously increasing impact on the environment, using about 50% of raw materials, consuming nearly 71% of electricity and 16% of water usage and producing 40% of the waste disposed of in landfills (1). Moreover, they are responsible for many harmful emissions, accounting for 50% of carbon dioxide (CO_2) emissions due to their operation, and an additional 18% caused indirectly by material exploitation and transportation (2). To mitigate the impact of buildings along their lifecycle, green building (GB) has emerged as a new building initiative during the last two decades. GB encourages the use of more environmentally friendly materials and the implementation of techniques to conserve resources, reduce waste generation and improve energy consumption, among others (3). This would result in environmental, financial and social benefits.

According to Anastaselos, Oxizidis (4), the potential for saving operational energy by applying GB methods is significant and can be estimated up to 40% of energy saving. Although there have been attempts to reduce energy use and carbon emissions by buildings, these efforts have yet to meet their fullest potential in designing and constructing high-performance buildings (5). This is mostly due to the lack of early decision-making in an integrated EED environment. It is widely believed that operational energy minimisation tasks should be made throughout a building project lifecycle (6). However, the most effective decisions related to energy efficient and sustainable design of a building facility are made in the early design and preconstruction stages (7). Early assessment of the performance of the buildings can help the Architecture, Engineering and Construction (AEC) stakeholders to choose better alternatives for the design to minimise energy consumption for a building lifecycle (8).

The conventional 2-Dimensional Computer-Aided Drawing (2D CAD) system requires investing large amounts of time for operational energy calculation. Therefore, using Building Information Modelling (BIM) can expedite this process and provide the opportunity to test and assess the impacts of different design alternatives and materials on the building (1). 'BIM is a digital representation of physical and functional characteristics of a facility' (9). According to BS1192, 'it is the management of information through the whole life cycle of a built asset, from

DOI: 10.1201/9781003207658-1

initial design through to construction, maintaining and finally de-commissioning, through the use of digital modelling' (10). Thus, decisions made in the early stages of design play a significant role in the level of sustainability throughout the life-cycle of the building (11). The ability to pinpoint the weaknesses of the design and implement changes based on the available alternatives can help the construction industry to mitigate the adversarial impact of building on the surrounding environment and enhance the greenness of building.

Hence, using BIM can positively manage and control the design process more meticulously and systematically (12). A BIM model represents the building as an integrated database of coordinated information from AEC disciplines. Beyond graphically depicting the design, much of the data needed for supporting sustainable design is captured naturally as the design of the project proceeds (13). In addition, the integration of BIM with performance analysis aspects greatly simplifies the often cumbersome and difficult analysis procedure. Finally, this approach can allow architects to obtain immediate feedback on design alternatives early in the design process (14).

1.2   BIM and Energy Efficient Design Problematisation

The current interface between the BIM and EED process is mostly user-driven (15). In other words, once the building is modelled and analysed, BIM does not have sufficient capacity to optimise operational energy based on the model. So, the user should, based on their knowledge and experience, either change the elements' properties and positions or play with the variables to get the optimum result in EED (16). The optimisation of the design is especially important when the sustainability performance of the building is required. As mentioned earlier, the BIM process can offer advantages in energy consumption estimation and optimisation, but experts do not effectively apply these potentials due to being user-driven and disjointed. In the literature, several statistics and simulation-driven methodologies have been developed to estimate and optimise energy consumption (17). Online building energy predictions based on artificial intelligence (AI) applications and web-based integrations can also be used in some areas (18).

Krygiel and Nies (19) suggested several innovations within BIM, such as improvements in software interoperability and integration of a carbon accounting tracker and weather data to provide the next steps in enhancing its capabilities with sustainability. Azhar (20) described the use of BIM to select building orientation, evaluate various skin options and perform daylight studies for its positioning on the chosen site during the design phase, thus enhancing its sustainability. Holness (21) noted that because of the trend in sustainability toward net-zero energy buildings and carbon emissions reduction, designers need to analyse the structure as a fully integrated dynamic design and construction process.

The current BIM can be referred to as *passive BIM* (22). However, *passive BIM* does not effectively support decision-making procedures. It cannot provide comprehensive sustainability analysis data such as energy consumption level and the design variables that need to be optimised. In contrast, BIM could be effectively transformed into *active BIM* because it is parametric and object-oriented (23). *Active BIM* is BIM that endorses intelligent decision-making platforms and can present the optimum values for the design variables to reduce the energy consumption of a building model. As a holistic view of the progress so far, it can be inferred that the lack of simple, integrated and practical decision-making procedures in BIM has remained unsolved. However, if this affliction is managed, it can assist academics, operators and designers to have an early perception of building energy performance and to propose solutions or alternatives for optimisation purposes.

Much research has been conducted on calculative, predictive, simulative, optimisation methods, building energy simulation and modelling. Yet, as will be elaborated in the next few chapters, these methods have not been practised or presented inbuilt within the integrated environment of BIM. One of the reasons lies in the lack of ideal interoperability between BIM and energy analysis software packages (24). Interoperability is the key to integrating BIM and external energy simulation, and its effective application enhances the inclusion and consistency of the information transition (25). The general unavailability of BIM vendor-neutral data formats and standards and issues regarding accessibility and security of data is another reason that the dynamic integration of BIM and EED has been hindered (26). The quality of data and its accessibility play a central role in BIM integration studies. Still, due to the proprietary nature of BIM applications, it is complicated to adopt the real time energy optimisation in BIM (27).

In addition, BIM-specific requirements are yet to be adequately embedded within the current state of the design phase. Thus, it creates disruptive possibilities in reaching the optimum EED settings (28). The need for more inputs and technical specifications in preparing models for BIM packages improves the semantic-rich content of BIM models, while 'the technology to collaborate on models has not yet delivered the industry requirements for BIM collaboration' (1). BIM can define an explicit and inbuilt configuration for digitised information of EED, but the current procedures are inefficient given the reasons mentioned previously. Thus, to proactively rectify building performance issues and improve energy efficiency, there is a need for robust methods and frameworks that can assist with detection, measurement and optimisation of energy performance (29). These methods need to be rapid, non-destructive, smart and integrated to be widely applied to building facilities.

As inferred from Figure 1.1 and discussed earlier in this chapter, today's EED in the built environment context confronts a wide range of challenges, such as less integrated platform, low interoperability between external energy simulation applications and lack of 'what if' scenarios analysis (1, 6, 20). In addition, the construction industry has faced the emergence of innovative methods of BIM and advanced data analytics like AI. AI is 'the science and engineering of making intelligent machines' (30), but these technological innovations heavily rely on integrated platforms to tackle the current challenges and issues of EED. These reasons point to the profound salience of AI-enabled active BIM for contemporary EED (see Figure 1.1).

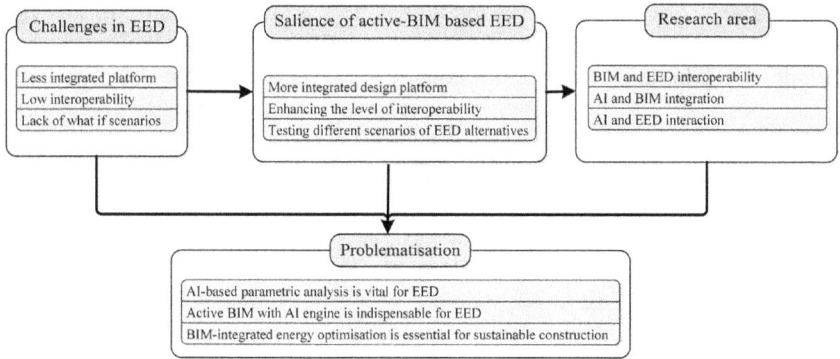

Figure 1.1 BIM-EED Problematisation

First and foremost, the introduction, inclusion and integration of AI into BIM and EED leads to energy efficient generative design implementation. Such an approach comprises automatic practices to develop different design alternatives that meet criteria like minimum energy consumption (31). Furthermore, it consists of algorithmic procedures driven by the immense working capacity of computer machines to generate a considerable number of alternative solutions that would otherwise be impossible to create (32). In essence, the idea is that various possible energy performances can correspond to its design for the same building facility. Hence, if alternative arrangements are created, these can be analysed to obtain the optimum performance through applying estimation approaches such as building energy simulation (33). Furthermore, the potential of each design solution for a possible improvement can be assessed by AI-based optimisation algorithms.

AI can greatly alleviate the problem of interoperability by enabling BIM to analyse energy consumption and optimise the design parameters within the BIM environment. The current post-design routine of energy analysis and optimisation using external software could be effectively shifted into the inbuilt EED in BIM (34). Furthermore, AI includes numerous categories of algorithms such as prediction, classification and optimisation, which present better flexibility in settings for EED (35). So, in line with the generative design remarks, it can pave the way for 'what if' scenario analysis for choosing various design parameters by applying different algorithms for prediction and optimisation of the operational energy of the building model in the design stage.

Moreover, the issue of consistency and homogeneity of the data can be mitigated, founded on the integrated platform of AI and BIM. AI enhances the level of information from non-optimised to optimised values based on the parametric capability (36). This capability is the key to keeping data homogeneity since AI could set the parametric attributes according to details in BIM (12). Hence, the non-proprietary semantics of information is fixed, which, in turn, eliminates the inaccessibility and unavailability of BIM vendor-neutral data formats problem.

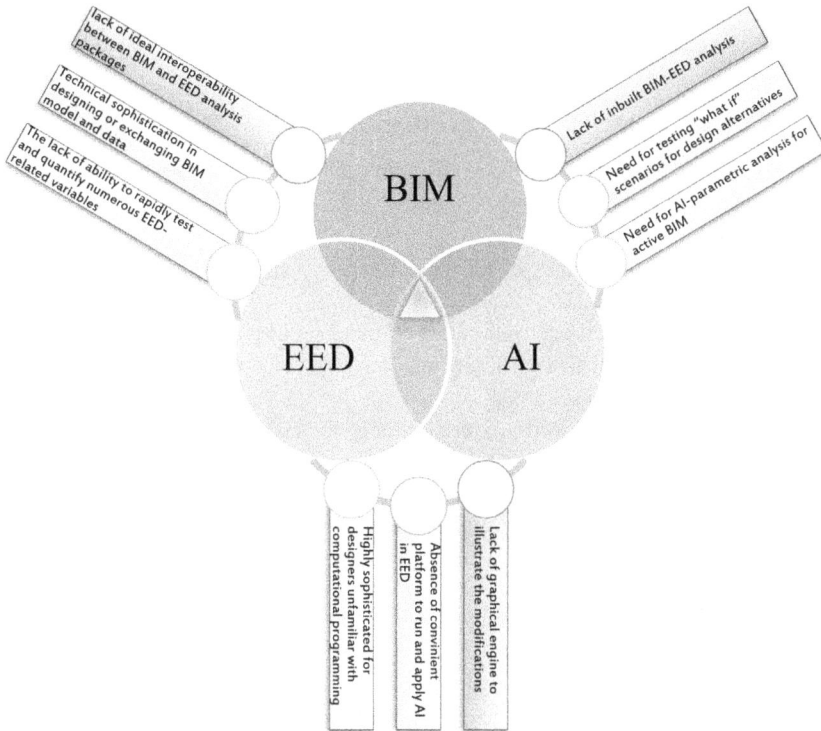

Figure 1.2 BIM-AI-EED Gap Diagram

Such potential further leads to the superior diffusion of BIM into EED in the interim of AI function in prediction, classification and optimisation of building design parameters. Figure 1.2 illustrates the summary of the gaps among three main domains of BIM, EED and AI.

All in all, the rationale for the present book is derived from the lack of knowledge on AI-enabled active BIM development and the paucity of studies in this respect in the construction industry. This is illustrated in Figures 1.1 and 1.2. In terms of problematisation, the literature review has established that developing the integration framework for BIM, AI and EED is central to establishing AI-enabled active BIM. However, this phenomenon continues to be an under-researched concept. In other words, a conspicuous lack of AI-based diligent BIM inquiry on EED within the extant literature was found. As a result, clarifying the major aspects of AI-enabled active BIM within the boundaries of the EED domain and within the built environment industry seems relevant and necessary. It should also be noted that there is a broad range of built environment facilities and their different characteristics, such as design types and layout, materials, construction technologies and priorities, in sustainable design. As such, it is almost impossible to develop a prototype that could cover all those criteria. Therefore, given their large share in worldwide energy usage and environmental impacts, stipulated by EIA (2006), residential buildings set the scope for the context of this book.

1.3   Objectives and Topical Investigation

This book is focused on a proper response to the following research and development questions in BIM, AI and EED topics:

1   What are the drawbacks of the current energy simulation and optimisation methods in buildings?
2   How is BIM capable of optimising the energy efficiency of buildings?
3   What are the important variables playing key roles in the energy consumption of residential buildings?
4   What types of AI algorithms are suitable for BIM to predict and optimise energy consumption?
5   How can an AI algorithm be linked to the BIM packages?
6   What is the process of optimisation for identified variables?
7   How can the body of knowledge and relevant stakeholders benefit from the developed BIM-AI-EED?
8   To what extent does the integration of BIM and AI deliver EED?

Therefore, this research monograph aims to develop an AI-based active BIM to optimise energy efficiency at an early design stage to obtain an initial estimate of energy consumption of residential buildings and optimise the estimated values through recommending changes in design elements and variables. Thus, the following specific objectives are sought:

1   Examining the potential and challenges of BIM to optimise energy efficiency in residential buildings.
2   Identifying variables that play key roles in energy consumption of residential buildings.
3   Investigating the AI-based algorithms in energy optimisation.
4   Developing a framework of AI application in BIM in terms of energy optimisation purposes and processes.
5   Assessing and validating the functionality of the framework using case studies.

1.4   Problem to Solution Discourse

Contextualising the problem to solve for this book implies that it commences with applying qualitative instruments to the exploratory tasks of objectives one and two, including examining the potential and challenges of BIM and EED and identifying the significant variables in residential energy consumption buildings. It is then actualised with quantitative instruments to address the third to the fifth objectives by modelling and simulating AI algorithms to investigate their functionality for energy optimisation, developing the integration framework of AI and BIM and validating the framework through case study verification.

Therefore, from the research design angle of this study, it can be realised that more than one research instrument is required to fulfil the following steps:

1   Identifying and prioritising the variables that play key roles in energy consumption of residential buildings.
2   Creating a comprehensive residential buildings dataset emphasising coverage of the full range of identified variables.
3   Developing AI-based algorithms using the collected dataset.
4   Linking the developed algorithms to BIM application.
5   Testing the workability of the framework and evaluating its performance.

Furthermore, the identified research steps should be scoped to apply the study development. Thus, the design stage is the focal point in the project lifecycle. This decision is made in light of the significant importance of the design stage in EED considerations. Henceforth, the construction and operation stages are excluded from the research. The developed framework and algorithms are designed for residential buildings only. Other building facilities are out of the scope, but the developed framework can be equally applicable to different buildings with modifications in parameters and dataset range. AI algorithms generally fall into two major functional categories of prediction and optimisation. So, two algorithms of artificial neural network (ANN) and decision tree (DT) for prediction and genetic algorithm (GA) for optimisation are applied. Last but not least, BIM includes a long list of tools from design to analysis, simulation and estimation. This study focuses on the Revit suite because of its popularity and parametric nature.

Extensive exploratory research must be constructed, considering the research scopes as the first step and qualitative. This is to identify the potential and challenges of BIM and EED and variables that play key roles in energy consumption of residential buildings by analysing the literature, creating a pool of variables and refining the list of variables. The qualitative method includes different instruments like the interview, focus group and Delphi (37). The Delphi method is selected here to complement the literature as it consists of an organised procedure in reaching a consensus among the respondents (38). Sourani and Sohail (39) confirmed that it also has an iterative nature and allows for running numerous rounds to achieve the outcome. In this study, experts conduct a three-round inquiry to brainstorm about the variables and synthesise action with literature, their prioritisation and confirmation of the final list of significant variables.

As the second step and from the quantitative side, the simulation method comes to the fore. This method enables researchers to analyse and investigate the eventual real influences or potential repercussions of various engineering situations and courses of action (40). Hence, a comprehensive residential building dataset simulated and analysed in BIM via considering identified variables should be provided and categorised by employing this method. This dataset should cover all variables from the literature and Delphi and have a sufficient amount of data for use in developing AI algorithms. In the last few decades, many researchers focused on several AI methods of optimisation algorithms. For the third step, the present study focuses on machine learning algorithms (41), including ANN and DT for data prediction and classification and GA for optimisation. Suitable and adaptable algorithms are established through data size reduction studies.

As the next step and by continuing the simulation instrument, the written algorithm should be linked to the BIM application based on the developed dataset in the previous step. The selected BIM application for this research is the Revit suite extensively used by BIM experts. It has a great potential for improvement because of its high level of interoperability with parametric algorithms (42). The developed algorithm is simulated in the Matlab software package and linked via Open Database Connectivity (ODBC) data exchange protocols to the Revit suite. Finally, an engineering-based case study instrument is applied to observe the developed framework's reliability and functionality for two conditions of pre- and post-optimisation. According to Algozzine and Hancock (43), a case study is an in-depth study of a particular situation. It is useful for testing whether scientific theories, models or frameworks work in the real world and its physical context. Thus, a comparative deviation report between the energy optimisation results of the active BIM and the previous record is made in order to test and validate the feasibility and suitability of the developed framework.

1.5   Significance of the Study

Enhancing BIM applicability in terms of sustainability and energy optimisation through developing AI-based active BIM is the main significant outcome of this research. There are many procedures, processes and platforms available now to model, simulate and optimise the energy consumption of buildings. Still, there is not a well-developed and integrated optimisation decision-making framework that can automatically estimate the energy consumption of buildings and then recommend specific changes for optimising purposes. The outcome of this study provides an early decision-making platform in an integrated EED environment emphasising the significant role of early stages of design on the sustainability of the building lifecycle. Furthermore, it can effectively shift the current practice of post-design energy analysis and optimisation to the inbuilt and unified EED in BIM through AI inclusion and operation. Accordingly, the problem of the less integrated platform and lower interoperability between external energy applications and BIM are mitigated.

In addition, the developed framework improves the semantic homogeneity and consistency of the information transfer in the EED. It aligns the parametric attributes of the AI and BIM model contents. As a result, the accessibility and availability of data are dramatically enhanced, and the issue of inconsistency of BIM vendor-neutral data formats problem is settled. Furthermore, the application of different batches of AI algorithms, including prediction, classification and optimisation, set the automatic capability of 'what if' scenarios for various design parameters of operational energy optimisation of the model. Therefore, a well-established framework of AI application in active BIM that gives an initial energy performance and introduces the optimised inputs for major design parameters is developed through this study. Ultimately, this research heads toward the higher diffusion levels of BIM and AI into the EED, which can be very useful for expanding the academic body of knowledge and its research and development effects on the industry.

In addition, this book sheds light on the prerequisites for developing this framework. A comprehensive literature review fully reveals the current state of the art, challenges, solutions, themes and gaps in the extant literature of EED, BIM and AI. The literature review and the three-round Delphi study led to the identification, prioritisation and confirmation of the significant factors in EED of residential buildings. Developing different batches of AI algorithms and using data size reduction techniques results in exploration of the most appropriate types of AI combinations to be utilised in the BIM platform.

1.6   The Outline

This research book consists of seven chapters, designed to cover details of the research from the rationale for doing the study to the methodology and data collection and analysis and, eventually, to presenting the discussion and conclusions.

Based on the book's background and justifications, Chapter 1 puts forward the classics of data-driven BIM for EED. It is further presented with the problematisation and problem to solution context. Chapter 2 focuses on an abridged account of the fundamental concepts and definitions associated with the topic. Acting as the backbone of the study, the theoretical framework of sustainability, information theory and optimisation are jointly discussed with their resulting technological implications for EED, BIM and AI. Chapter 3 maps out the upstream concepts of Chapter 2 to a more detailed focus on the critical appraisal of BIM and EED and deploying gap-spotting techniques of the body of the knowledge; Chapter 4 addresses building energy parameters, their geometrical, topological and semantic associations and the prioritised variables for residential building energy simulations. Chapter 5 aims to analyse the application of AI in energy efficient building development and its data generation, prediction and optimisation aspects. Chapter 6 intends to indicate the procedural steps of developing the framework of BIM and AI integration concerning EED. It further verifies the applicability and functionality of the framework in the built environment context by applying the case study approach. Finally, Chapter 7 presents BIM and the data-focused visionary leap for future building design. It concludes the book by summarising the research aim and objectives, discussing the contributions to the body of the knowledge, implications for practice, the current limitations and recommendations for the future of the topic.

References

1 Oduyemi O, Oduyemi O, Okoroh MI, Okoroh MI, Fajana OS, Fajana OS. The application and barriers of BIM in sustainable building design. Journal of Facilities Management. 2017;15(1):15–34.

2 Kibert CJ. Sustainable construction: Green building design and delivery: John Wiley & Sons; 2016.

3 Reddy BV, Jagadish K. Embodied energy of common and alternative building materials and technologies. Energy and Buildings. 2003;35(2):129–37.

4 Anastaselos D, Oxizidis S, Manoudis A, Papadopoulos AM. Environmental performance of energy systems of residential buildings: Toward sustainable communities. Sustainable Cities and Society. 2016;20:96–108.

5 UNEP. SBCI (Sustainable Building and Climate Initiative). Vision for sustainability on building and construction: United Nation Environment Program; 2014. Available from: www.unep.org/sbci/AboutSBCI/annual_reports.asp.

6 Bynum P, Issa R, Olbina S. Building information modeling in support of sustainable design and construction. Journal of Construction Engineering and Management. 2013;139(1):24–34.

7 Basbagill J, Flager F, Lepech M, Fischer M. Application of life-cycle assessment to early stage building design for reduced embodied environmental impacts. Building and Environment. 2013;60:81–92.

8 Ekici BB, Aksoy UT. Prediction of building energy needs in early stage of design by using ANFIS. Expert Systems with Applications. 2011;38(5):5352–8.

9 Eastman C, Panushev I, Sacks R, Aram S, Venugopal M, See R, et al. A duide for development and preparation of a national BIM exchange standard: PCI-Charles Pankow Foundation; 2011.

10 BS1192. 1192: 2007: Collaborative production of architectural, engineering and construction information, code of practice. ISBN: 978 0 580 58556 2. British Standards Institute (BSI); 2008.

11 Attia S, Gratia E, De Herde A, Hensen JL. Simulation-based decision support tool for early stages of zero-energy building design. Energy and Buildings. 2012;49:2–15.

12 Noack F, Katranuschkov P, Scherer R, Dimitriou V, Firth S, Hassan T, et al., editors. Technical challenges and approaches to transfer building information models to building energy. eWork and eBusiness in Architecture, Engineering and Construction: ECPPM 2016: Proceedings of the 11th European Conference on Product and Process Modelling (ECPPM 2016), Limassol, Cyprus: CRC Press, September 7–9 2016; 2017.

13 Cerovsek T. A review and outlook for a 'Building Information Model' (BIM): A multi-standpoint framework for technological development. Advanced Engineering Informatics. 2011;25(2):224–44.

14 Azhar S. Building Information Modeling (BIM): Trends, benefits, risks, and challenges for the AEC industry. Leadership and Management in Engineering. 2011;11(3): 241–52.

15 Wu W, Issa R. BIM execution planning in green building projects: LEED as a use case. Journal of Management in Engineering. 2014;31(1):A4014007.

16 Wang J, Li J, Chen X, editors. Parametric design based on building information modeling for sustainable buildings. Challenges in Environmental Science and Computer Engineering (CESCE), 2010 International Conference on: IEEE; 2010.

17 Machairas V, Tsangrassoulis A, Axarli K. Algorithms for optimization of building design: A review. Renewable and Sustainable Energy Reviews. 2014;31:101–12.

18 Foucquier A, Robert S, Suard F, Stéphan L, Jay A. State of the art in building modelling and energy performances prediction: A review. Renewable and Sustainable Energy Reviews. 2013;23:272–88.

19 Krygiel E, Nies B. Green BIM: Successful sustainable design with building information modeling: John Wiley & Sons; 2008.

20 Azhar S, Carlton WA, Olsen D, Ahmad I. Building information modeling for sustainable design and LEED® rating analysis. Automation in Construction. 2011;20(2):217–24.

21 Holness GVR. BIM: Gaining momentum. ASHRAE Journal. 2008;50(6):28–40.

22 Welle B, Haymaker J, Rogers Z, editors. ThermalOpt: A methodology for automated BIM-based multidisciplinary thermal simulation for use in optimization environments. Building Simulation. 2011;4:293–313.

23 Moon H, Kim H, Kamat V, Kang L. BIM-based construction scheduling method using optimization theory for reducing activity overlaps. Journal of Computing in Civil Engineering. 2013:04014048.

24 Sanguinetti P, Abdelmohsen S, Lee J, Lee J, Sheward H, Eastman C. General system architecture for BIM: An integrated approach for design and analysis. Advanced Engineering Informatics. 2012;26(2):317–33.

25 Woo J-H, Menassa C. Virtual Retrofit Model for aging commercial buildings in a smart grid environment. Energy and Buildings. 2014;80(0):424–35.

26 Gray M, Gray J, Teo M, Chi S, Cheung YKF. Building information modelling: An international survey. 2013.

27 Volk R, Stengel J, Schultmann F. Building Information Modeling (BIM) for existing buildings: Literature review and future needs. Automation in Construction. 2014;38:109–27.

28 Sawhney A, Singhal P. Drivers and barriers to the use of building information modelling in India. International Journal of 3-D Information Modeling (IJ3DIM). 2013;2(3):46–63.

29 Ham Y, Golparvar-Fard M. EPAR: Energy performance augmented reality models for identification of building energy performance deviations between actual measurements and simulation results. Energy and Buildings. 2013;63:15–28.

30 McCarthy J, Hayes P. Some philosophical problems from the standpoint of artificial intelligence: Stanford University USA; 1968.

31 Singh V, Gu N. Towards an integrated generative design framework. Design Studies. 2012;33(2):185–207.

32 Chakrabarti A, Shea K, Stone R, Cagan J, Campbell M, Hernandez NV, et al. Computer-based design synthesis research: An overview. Journal of Computing and Information Science in Engineering. 2011;11(2):021003.

33 Soares N, Bastos J, Pereira LD, Soares A, Amaral A, Asadi E, et al. A review on current advances in the energy and environmental performance of buildings towards a more sustainable built environment. Renewable and Sustainable Energy Reviews. 2017;77:845–60.

34 Banihashemi S, Ding G, Wang J, editors. Developing a framework of artificial intelligence application for delivering energy efficient buildings through active BIM. COBRA 2015; 2015: RICS.

35 Yang C, Li H, Rezgui Y, Petri I, Yuce B, Chen B, et al. High throughput computing based distributed genetic algorithm for building energy consumption optimization. Energy and Buildings. 2014;76:92–101.

36 Tao F, Zhang L, Laili Y. Configurable intelligent optimization algorithm: Springer; 2014.

37 Ritchie J, Lewis J, Nicholls CM, Ormston R. Qualitative research practice: A guide for social science students and researchers: Sage; 2013.

38 Hsu C-C, Sandford BA. The Delphi technique: Making sense of consensus. Practical Assessment, Research & Evaluation. 2007;12(10):1–8.

39 Sourani A, Sohail M. The Delphi method: Review and use in construction management research. International Journal of Construction Education and Research. 2015;11(1):54–76.

40 Prive NC, Errico RM. Observing system simulation experiments: An overview; 2016.

41 Engelbrecht AP. Computational intelligence: An introduction: John Wiley & Sons; 2007.

42 Kensek K, Noble D. Building information modeling: BIM in current and future practice: John Wiley & Sons; 2014.

43 Algozzine B, Hancock D. Doing case study research: A practical guide for beginning researchers: Teachers College Press; 2016.

2   Sustainability, Information and Optimisation

Antecedents of the Data and BIM-Enabled EED

2.1   Paradigms of Sustainability, Information and Optimisation Theories

2.1.1   Sustainability

The current environmental crisis threatens human survival; population growth has led to excessive urbanisation and decreased the capacity of lands to support (1). Developed and developing countries hugely utilise natural resources and dig to find other sources that may be intact (2). The rapid depletion of these resources, global warming, climate change and runway economic and industrial expansions have sparked serious debates over the definition and implementation of sustainability. Although it is found problematic to reach an international agreement regarding the theory of sustainability, it has been recognised that the ongoing expansion of humanistic activities leads to the endangering of our ecosystem (3).

In this respect, sustainable development principles have been stipulated through many local, regional and global policies and applied in industrial and commercial sectors. Around 50 years have elapsed from a campaign of the environmental movement during the General Assembly of the United Nations in 1972 through nominating World Environment Day. There seems to be an increasing responsibility to shift from unsustainable paradigms in development to more sustainable ones (4).

Sustainability is one of the most difficult concepts science has confronted so far. World Commission has defined two popular statements for manifesting this concept on Environment and Development and the National Research Council. The former says 'development that meets the needs of the present without compromising the ability of future generations to meet their own needs' (5), and the latter states 'the reconciliation of society's development goals with the planet's environmental limits over the long term' (6). Both definitions are process-oriented and attempt to address environmental, social and economic principles. However, Brundtland's description is the most established and can be applied and generalised at local, regional, national and international levels (7). This definition forms, therefore, the base for this research.

In the past four decades, because of a constant state of sustainability and sustainable development, new fields of science, information and philosophies have been

DOI: 10.1201/9781003207658-2

added to the body of knowledge (8). Furthermore, different creative and scientific endeavours have been practised, including sustainability assessment indicators and models, for responding to the tremendous risks imposed by climate change and global warming imperatives (9). Breaking this body of knowledge into inter-disciplinary collaborations of research can be an important part of enhancing the understanding of sustainability in society (10).

Sustainability needs to be embodied in the subtexts of different areas of knowl-edge ranging from engineering, physics and applied sciences to the social sciences and economy. But, unfortunately, this irresistible momentum cannot be maintained due to the absence of an underpinning theory for sustainability that leads to many ambiguities (11). For example, we do not exactly know how to measure the sus-tainability level accurately or what appropriate metrics are for covering a wide range of issues like global warming and climate change (12). Additionally, the cri-teria for identifying a sustainable procedure from an unsustainable one in cleaner production and activities such as resource minimisation, improved eco-efficiency and source reduction and their underlying foundations are yet to be fully recog-nised (4).

Looking for methodologies and approaches that permit transdisciplinarity is necessary in order to investigate what happens between, across and beyond disci-plines (13). Nicolescu (14) states that 'transdisciplinarity concerns the dynamics engendered by the simultaneous action of several levels of reality'. This capability is certainly required for the sustainability issue to develop knowledge at numer-ous levels while keeping a holistic view of the world (15). Efforts of modelling and evaluating sustainability are among the actions that can positively contribute toward this task (16).

Information theory and its modern paradigms, prototypes and processes can establish a common ground for developing frameworks of sustainable develop-ment (17). Also, the advanced state of this theory, coupled with the opportunities that information technology (IT), cybernetics and informatics provide us with, can be used to develop an integrated model for global issues.

2.1.2   *Information Theory*

The concept of entropy of random variables and procedures set a stepping stone of information theory as 'a conceptualisation and characterisation of processes that allows for storing and communicating of data' (18). Information theory includes three major paradigms – cybernetics, cognitive and informatics – in which data are carriers of meaningful content or messages for the related recipient. This is a typi-cal approach of cybernetics that has been understood so far as generic information theory. This approach is applied to create a comprehensive model describing the characteristics and attributes of information in cybernetics (19).

The method of data transmission between systems of the same nature, intro-duced by Miller and Miller (20) in their seminal study as technical, social and techno-social systems, is called the information process. It is identified in cyber-netics by creating a signal storing and transmitting data. Consequently, information is considered a particular signal that preserves data content as a message. That is to say, in cybernetics:

- Information process ≡ generating a signal
- Information ≡ data containing a message (17)

Developing informatics and IT, conceptually and operationally, has been enormously influenced by cybernetics notions generated from information theory. However, some questions regarding the nature and meaning of data cannot be answered through cybernetics, and these are better explained with the semantics paradigm (17).

Semantics is a critical factor in interpreting the meaning and nature of information in cognitive theory. Information processes are defined as reflecting objects on human beings' consciousness (21). Semantics is also coupled with developments in the cognitive approach to cultivate AI and knowledge-based information systems (IS) (22). Equating information with new knowledge, achieved from the cognition mechanism, accurately describes the cognitive paradigm. In other words:

- Information process ≡ cognitive process
- Information ≡ new knowledge

This cognitive paradigm deals with information quite differently compared to cybernetics. It mostly focuses on the subjective and cognitive rather than the objective and physical aspects of information. On the other hand, both define information processes as the theoretical foundation for formalising and characterising information (17).

The third paradigm, informatics, has a more practical and technical nature utilised in conjunction with information theory. From this point of view, information processes primarily provide information for managing technologies and maximising their effectiveness. As a scientific discipline, informatics is founded upon cybernetics and extended to cover the practical implications of information, but it does not distinguish data from information. Moreover, in line with the cognitive paradigm, it incorporates human agents as a necessary parameter in the information process (23). Such a pragmatic paradigm can be represented as:

- Information process ≡ storing, transmitting and processing of data
- Information ≡ data process for a specific aim

The paradigms mentioned earlier pinpoint particular facets of information, but all of them deal with the same stuff: data (Table 2.1). All kinds of data can convert to information, disregarding their disciplinary records, but it is hard to acquire the understanding of the optimal level of information generation for modelling sustainability (24). However, these paradigms emphasise transdisciplinarity to portray an accurate picture of reality (11). Among the three discussed paradigms of information theory, informatics is chosen for this research because of its pragmatic and technical approach toward the role of information in managing and maximising

Table 2.1 Information Theory Paradigms

Paradigm	Information Process	Information Meaning	Interpretation
Cybernetics	Generating a signal	Data containing a message	Meaningful signal content (information) preserving data characteristics as a message
Cognitive	Cognitive process	New knowledge	Equating information with new knowledge achieved from a cognition mechanism
Informatics	Storing, transmitting and processing of data	Data process for a specific aim	Providing information for managing the technologies and maximising their effectiveness

Source: Adapted from Todorov & Marinova (2010)

the effectiveness of sustainability. Informatics paves the way of the information process, storing, transmitting and processing data to maximise the effectiveness of sustainability (17). But, to analyse the modelling competency of informatics and its relevancy to sustainability, there is a need to study optimisation as a cohesive unit for information theory serving a robust interdisciplinary bridge.

2.1.3   Optimisation Theory

Optimisation is a branch of knowledge extensively employed in engineering, applied sciences, social sciences and other fields. It deals with choosing the optimal solutions within different problems, introducing computational techniques to find the best decisions and exploring their calculus performance (25). The terminology of 'optimise' is generally used for substituting for the words of maximise or minimise, and its basic principle is to identify the best choice for given alternatives. To do so, the entire possible decision should be tested, and the authenticity of the optimal choice should be validated (26).

Any optimisation problem requires an objective function that generally contains more than one variable for optimisation (27). Of course, a single variable function can also be created. Still, from an optimisation perspective, there would not be room for improvement in the optimisation results because it quickly reaches the end without fully analysing the whole choices. In addition, objective functions including multiple functions are referred to as multi-objective optimisation, which usually have complex interrelationships and take time to optimise. However, these types are more precise in delivering the optimisation results (28).

Concerning the character of the optimisation purpose, three types of variables, including real, integer and a combination of both, can be mentioned. For real (continuous) variable problems, experts usually try to find a group of genuine numbers (29). On the other hand, integers, sets or graphs from finite or infinite sets of objects are investigated for discrete variables. The third type combines continuous and discrete variables and is termed combinatorial optimisation. These problems usually need completely different optimisation and problem-solving techniques (25).

Constrained and unconstrained are another optimisation problem classification. Nevertheless, this classification is a bit questionable as some experts believe that there are no unconstrained problems throughout the world and that all optimisation problems do indeed have constraints (30). Notwithstanding this critique, this category seems important within optimisation theory because of the opportunity in converting constrained to a chain of unconstrained problems (29). Linearity or nonlinearity is the other significant attribute of optimisation problems, in which nonlinear models need more complicated and advanced approaches in optimisation (28).

Functions can also be broken into convex or non-convex and differentiable or non-differentiable parts (25). Several optimisation methods are devised assuming that the function curves outward, and that is why convexity is one of the major elements of classical optimisation theories. Finally, differentiability is the third subgroup of function classifications and is mostly applicable for continuous functions. It depends on derivative-based algorithms and cannot be implemented for derivative-free functions (31). The optimisation paradigms can be generally classified as shown in Figure 2.1.

Given the overview of optimisation theory and its principles, it is important to link this theory to the ultimate aim of this study. The reason for applying optimisation theory lies in assisting information theory and its pragmatic driver, informatics, to reach an enhanced level of sustainability. The unconstrained multiple objective functions consisting of mixed variables of continuous and integer could convex the information contained in the informatics toward the optimised level. Therefore, informatics employs optimised information to run the information process and maximise sustainability.

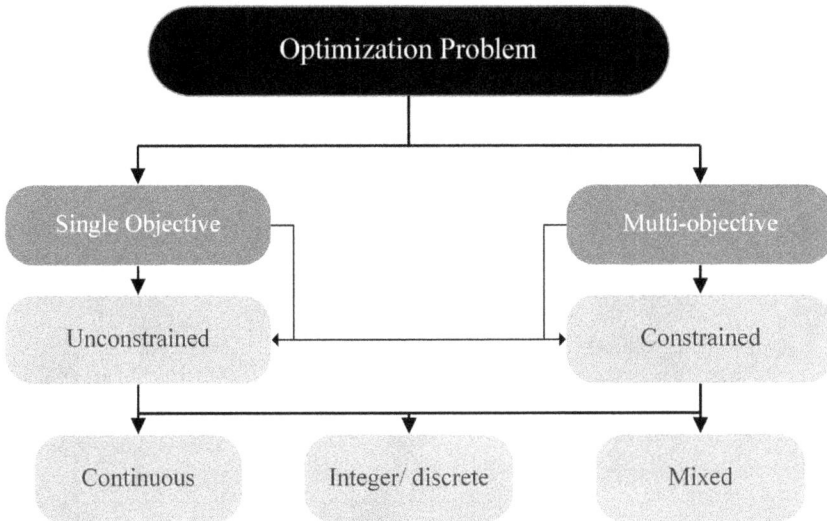

Figure 2.1 Classifications of Optimisation Paradigms
Source: Adapted from (25, 31, 32)

2.1.4   Trilateral Interaction

This section presents the interaction of three theories of sustainability, information and optimisation as the underlying theme for addressing this book and provides a schematic view on the conceptual model. The interaction and integration of these three theories establish a theoretical foundation for this study. For modelling this interaction, these systems must be centrally depicted. Following the literature of their foundations, Figure 2.2 indicates the three domains of sustainability, information and optimisation theories as the main operating components of this model.

Reaching an integration for this upstream will reveal uncharted territory and may even make it an insoluble enigma. For instance, what the agreeable level of embracement is that these domains share. Moreover, the present interaction of sustainability, information and optimisation provokes some arguments that this study should address.

The first argument regards the character of interactions and the type of the model among these theories. Current tools such as econometrics, environmental and sociometric models for sustainability, informatics for information theory and iterative, algorithmic and calculus techniques for optimisation theory can shape the nature of interactions within each approach (33). Nonetheless, a distinctive model dealing with all major elements and new advanced technologies would be required to model this meta-system. Moreover, it is axiomatic that these paradigms continuously change in and shape different interplays.

The second argument investigates the nature of the information obtained from the interaction. The released information should be treated as a system status relevant to generating, transmitting and receiving information items and analysed as

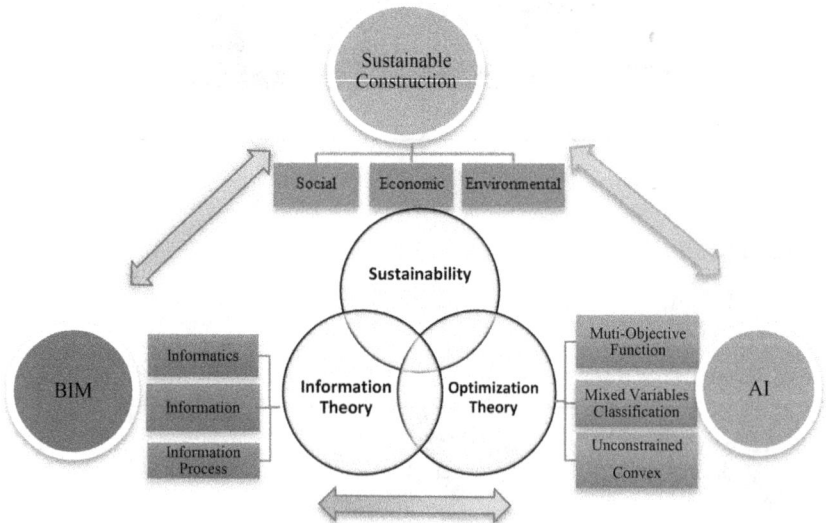

Figure 2.2 Interaction of Three Theories and Their Components

practical informatics. These procedures also transform, share and optimise these status collections, whereas the sustainability derivatives dictate these processes' course, approach and content. Therefore, achieving sustainability makes it imperative for regular generation, transformation, transmission, receiving and optimising information concerning the target (34).

Sustainable development has been becoming a global phenomenon strengthening the links and synchronising the movements among humanity, nature and the economy (33). Along with the growth of interest of policy makers and the imposition of international environmental regulations, the need for environmentally and ecologically related information is perceived more than ever (34). This procedure is a cycle of continuous development and adjustment. It indicates the interaction of sustainability, information and optimisation theories, which requires continuous supervision and analysis to find an intelligent decision. Cutting-edge scientific discoveries in information theory and computer-aided algorithms of optimisation theory can facilitate managing the scale and spectrum of sustainability issues and allow for running this framework properly (17).

'Information is central to guidance, and guidance for a sustainability transition needs information on both where we want to go as well as how well we are doing at getting there' (35). What is claimed here is that information can offer such guidance and feed the transdisciplinarity pillars if the optimised information derives through the power of intelligent agents during a sustainable phase. A journey should be set from the upstream to the downstream to bring the interaction idea of the three theories of sustainability, information and optimisation into a practical context. This journey conceptualises the relevant derivatives to reach harmony in the system as system integration is one of the high-frequency terms reflecting the concerns of recognising the interconnections among different components for sustainability. This is especially a prominent finding that addresses a need for system thinking to elucidate this journey.

It should be noted that the evolution of data and its effective integration is the key problem in reaching a harmonised system. As to the informatics processes of information theory, well-structured data could be conceptualised as the common data environment (CDE), a harmonic place for collecting, managing and sharing information among a team working on a system (36). CDE is the single source of information used to collect, manage and disseminate documentation, the graphical model and non-graphical data for the whole system (i.e. all project information, whether created in a BIM environment or a conventional data format). Creating this single source of information facilitates collaboration between system members and helps avoid duplication and mistakes (37).

Science and cutting-edge technologies such as BIM and AI can play a crucial role here. This knowledge helps construct understandable foundations to reach a sustainable system through preserving harmony among various domains (17). It can compose or combine several separate branches of expertise or fields to draw appropriately from multiple disciplines to redefine problems outside of normal boundaries. Finally, it is able to reach solutions based on a new understanding of complex situations (38). So, an inter- and multidisciplinary approach is necessary to link the different levels of this study to determine the hierarchy of subsystems and direct, indirect and cross-interactions among the criteria.

Based on Figure 2.2, the described interaction process among three fundamental theories of sustainability, information and optimisation is regarded as the core of this integration system. It presents information theory and its functional driver; informatics works toward sustainability's information process. This batch of information is being optimised through the optimisation paradigms via setting objective functions, mixing continuous and integer variables and convex tasks for reaching the maximised effectiveness of sustainability. Therefore, such a theoretical integration sets the journey to the downstream and practical level, which should be elaborated in the construction field to translate the discourse into practical actions. Hence, sustainability is narrowed down into sustainable construction drivers. BIM, then, comes to the fore as the representative of informatics applications and CDE driver in the construction industry (39).

BIM stores, transmits and processes the optimised information, which is optimised through AI as the practical machine of optimisation theory paradigms (40). According to BS1192 (36), CDE is the 'collaborative production of architectural, engineering and construction information and code of practice' containing various information environments. It may comprise a supply-side CDE used by the project delivery team and the information environment that facilitates an employer-side document and data management system for the receipt, validation and approval of project information. Effective use of a CDE builds an accurate and well-structured dataset as the stages progress. Through this setting, the integration of three fundamental theories of sustainability, information and optimisation is linked to their three functional drivers of sustainable construction, BIM and AI and establishes a theoretical framework for this research. For this reason, in the next sections, sustainable building, BIM and AI are elaborated, and their focus on EED is justified.

2.2   Sustainable Construction Drivers

Extensive research has proven that among the myriad of industries, civil works consume 60% of the raw materials obtained from the earth, of which the building sector contributes to 40% of this damage (41). Likewise, greenhouse gas (GHG) emitted from buildings to the environment constitutes approximately 40% of total world GHG emissions (42). To give a full picture of this concern as associated with construction, it is worth mentioning that among the top ten worst pollution problems in the world for the decade, *urban air quality* and *indoor air pollution* are directly, and *contaminated surface water* and *industrial mining activity* are indirectly, related to this industry (43).

The need for environmentally friendly construction practices led to the advent of a movement known as green building (GB). GB has been developed to encourage use of less toxic materials, applying more resource conservation techniques and improving energy usage patterns, among others (44). 'GB refers to both a structure and the application of processes that are environmentally responsible and resource-efficient throughout a building's lifecycle: from planning to design, construction, operation, maintenance, renovation, and demolition' (45). The ultimate goal of GB is to eliminate the negative impacts of building on its surroundings and human health. This would result in environmental, financial, economic and social benefits (46).

A study published by McKinsey Company's report on sustainability and resource productivity (47) shows that GHG emissions of GHGs can be reduced by 6 billion tonnes per year by implementing GB. Another study shows that well-designed sustainable dwellings compared to conventional ones consume, on average, 25% less water and conserve energy up to 45% (48). During the last decade, many research projects, studies and tools have been carried out concerning tackling sustainability issues in construction. Holistically, these works mainly focus on (49):

- GB assessment tools
- GB process implementation
- Energy and carbon footprint reduction
- Built environment hydrologic cycle
- Sustainable material loops
- Waste reduction
- Indoor air quality
- Sustainable site and landscape
- Sustainable technologies in green building (Figure 2.3)

Figure 2.3 A Sustainable House in Eagle Rock, California.

'The house makes use of a variety of green technologies and materials including a solar energy system, rainwater collection, grey water recycling, passive thermal rock wall, poured in place concrete (countertops, tubs, and sinks) with 50% recycled fly ash, recycled wood ceiling' (50).[1]

However, these efforts have yet to meet their fullest potential in managing the sustainability of high-performance buildings. According to the annual report of the Sustainable Building and Climate Initiative (SBCI) of the United Nations Environment Program (UNEP), the performance of buildings is 'far below current efficiency potentials' (51). They still are the biggest contributors to GHG emissions and consumers of the world's electricity. Numerous barriers hinder the implementation of sustainable buildings (52):

- Financial disincentives
- Lack of whole system thinking
- Construction process complexity
- Lack of communication and coordination among parties involved
- Inadequate research studies
- Unavailability of novel methods and tools
- Lack of integrated design methods

At a glance at these barriers, it is obvious that they fall into two major categories: organisational and technical. So, sustainable construction, to be effectively implemented, requires overcoming such barriers (53). There are multiple strategies to follow (either through reducing consumption, increasing efficiency or finding less adverse methods), but in all conditions, innovations are imperative (54). This process requires agents to promote innovations through coordination, networking and communication (55). Nevertheless, the broad content of sustainable construction and the plurality of the meanings of GB make it hard for experts to reach a consensus in concept. It may even cause a disagreement in problem formulations and lead to contradictory solutions (56). This fact hampers cooperation, which in turn inhibits developing innovative solutions (57).

Banihashemi, Hosseini (58) criticised the absence of a connection between innovation and sustainable construction. Construction experts typically have to innovate to improve performances and meet the increasingly complex and fast-changing needs for sustained quality of the built environment (54). But, of course, as Yang (55) argued, sustainability is currently no more than a breakeven point, and it is necessary to change to sustainable and then restorative and regenerative construction (59). He called this process a continual improvement in structure. So, conventional mono-disciplinary innovations can no longer modify the technology regime and perceptions of underlying concepts of current construction methods (60).

Furthermore, construction is one of the most information-dependent industries, mainly due to its extended fragmentation (61). Construction projects are often complex and unique, involve many activities and require the employment of several human resources with various specialisations (58). Thus, the amount of information generated and exchanged during the construction process is enormous, even for small-sized projects (62). For this reason, Sodagar and Fieldson (63) state that having efficient access to the information is key for facilitating integrated design methods, a collaborative construction process, analysing their sustainability and so on, making construction more sustainable. GB is prevented from deficiencies in communication among parties involved, whereas fluent cooperation and networking are essential to acquire momentum in the design, construction and operation of buildings (64). The efficient use of information and the active cooperation of parties involved via the management and sharing of information methods can contribute to sustainable construction. This occurs through overcoming the obstacles of capturing and managing the data and information and the technical-organisational barriers (65).

The absence of systematic information management has forced functional disciplines to follow their own decisions. These decisions are mostly without considering the impacts on other fields that provoke clashes in coordination and communication in construction projects and impede sustainability drivers (66). At the same time, efficient information management can enable experts to consider the wide range of sustainable building components, including building performance, lifecycle costs, a rapid adaptation of design and environmental impacts (67). In contrast, knowledge sharing is also considered problematic for sustainable construction, especially when different sources of information are available. Investigating the other sources of sustainability-related details for the Swedish construction industry indicated that providing biased or distorted information can be one of the possible risks in this regard (56).

All in all, despite serious attempts, construction is deficient in tools (1) to implement GB in all stages and (2) to contrast different alternatives and methods. It is vital to provide tools to be objective, reliable and accessible (68). That is why in recent years, various innovative methods have been introduced to converge technologies and knowledge from different disciplines and scientific theories and to eliminate GB's technical or organisational barriers. Information and communication technology (ICT) is one of these new methods being utilised in answering the critiques laid out earlier, through providing an integrated chain of software and hardware. 'Using new information technologies gained and exchanged knowledge and experience shall become instrumental to the success of approaching sustainable buildings and sustainability generally' (69). The impressive use of such ICT methods and mechanisms requires integrating tools for the designing, construction and operation phases and the development of intelligent methods for information management (68). As a specially designed process and innovative approach, BIM produces parametric modelling and dynamically presents building information that can significantly contribute to sustainable construction (70).

2.3   BIM and Sustainable Construction

BIM represents the building as an integrated database of coordinated information. Beyond graphically depicting the design, much of the data needed for supporting sustainable building is captured naturally as the design of the project proceeds (71). It paves the way for superimposing multidisciplinary information in an integrated model, which creates an opportunity for measuring the sustainability state of the project throughout its lifecycle (72). According to the seminal book *Green BIM* by Krygiel and Nies (73), BIM has the potential to aid sustainable design through:

- Selecting a suitable building orientation
- Analysing daylighting
- Reducing water needs in a building
- Reducing energy consumption through integration with energy analysis software
- Featuring renewable energy potential
- Reducing material needs
- Decreasing waste and carbon emissions by logistic site management

Architects can import data into the BIM model to geographically locate the project and present the climate, location and surrounding area. The model can then be reorientated and edited on-site based on the real coordination for reducing resources and efficiently exposing to solar radiation (74). BIM can analyse the mass and form of models for enveloping analysis and window to wall ratio. Likewise, engineers utilise BIM to decrease energy consumption by exporting the 3D model to a particular energy analysis software and calculating light reflectance and transmittance (75). Effectively coordinating logistics through site analysis and modelling, including wetlands and protected habitats, can help contractors eliminate possible issues. It even easily quantifies the amount of materials extraction of buildings to simplify reusing or recycling measures (76). Subcontractors can also use BIM to reduce waste and integrate supply chain data to decrease carbon emissions. Finally, the BIM file arranges an array of required resources to advise the project team regarding sustainability issues from the incipient stages of a project (77).

In recent years, scholars have done extensive research on BIM and GB design and construction potentials from different points of view (Table 2.2):

- BIM for modelling and analysing of building performance, including energy and thermal simulation (78–81), lighting simulation (82, 83), daylighting simulation (84–86) and waste (87, 88)
- BIM for optimising an integrated design process (89–91)
- BIM for evaluating sustainability of buildings concerning leadership in energy and environmental design (LEED) and other environmental assessment tools (72, 92–95)

With technological advancement and wider use of BIM, its integration with sustainable design applications have become ubiquitous. For instance, Inyim, Rivera and Zhu (97) developed a BIM-based tool enabling decision-making for the sustainability of the design phase. Clevenger and Khan (98) analysed the contribution of BIM from design to fabrication processes for building materials. They contended that building delivery performance could be improved, and

Table 2.2 Literature Examples of BIM and Sustainable Design Archetypes

No.	Reference	BIM & Building Performance		BIM & Integrated Design	BIM & Green Certification
		Thermal	Lighting		
1	(96)	✓			
2	(81)	✓			
3	(80)	✓			
4	(84)		✓		
5	(83)		✓		
6	(85)	✓	✓		
7	(79)	✓			
8	(78)	✓			
9	(91)			✓	
10	(90)	✓		✓	
11	(76)			✓	
12	(95)				✓
13	(72)				✓
14	(93)				✓
15	(94)				✓
16	(92)				✓

unnecessary environmental impacts can be mitigated by reducing design errors and any miscommunications via BIM functions. Biswas and Wang (99) established a BIM-extended tool to incorporate BIM technology in evaluating environmental consequences from design decisions. This study could be regarded as one of the earliest attempts in BIM employment for the rating and certification of GB. In the same vein, Gandhi and Jupp (100) evaluated the potential utilisation of BIM for the Australian Green Star Building certification.

Apart from the theoretical implications of BIM for GB, Azhar (101) investigated the applicability of this tool through a survey among 145 experts of AEC within the US. First, concerning the time saved in applying BIM-based sustainability analysis vis-à-vis traditional analysis, 54% of respondents recognised some time savings, and 23% declared significant time savings. Furthermore, concerning monetary savings, similar to the first question, 51% of respondents asserted that they are experiencing some cost savings, and 26% realised significant cost savings. In a nutshell, it was observed that implementing BIM-based sustainability analysis resulted in 'satisfaction' up to a certain degree as regards the traditional approach.

Moreover, Malkin (102) conducted multiple case studies in the design and construction stages of large-scale projects in Australia and demonstrated the positive role of BIM in making these projects more sustainable. They effectively applied Revit package together with Ecotect Analysis to keep the overall concrete fly-ash volume up to 20% and meet the sustainability standards.

But there is a long way to go to implement BIM in sustainable design fully. BIM is still in its infancy, and attaining the desired quantum leap entails radical reorientation in the current state. The main reasons for this are (49):

- The lack of ideal interoperability between BIM and analysis software packages
- The need for more inputs and technical specifications in the BIM model
- The need for integration of carbon accounting data
- The information lost in the data translation between different parties
- The lack of ability to rapidly test and quantify numerous sustainability-related variables

These factors envision the core value of information as the bottom line of flawless BIM-based sustainability analysis. Bynum, Issa and Olbina (103) conducted a survey among 123 pioneer companies of AEC in the US; around 60% of respondents stated that 'BIM, as a multidisciplinary tool, currently does not operate in an optimal standardised format, making it difficult to translate data seamlessly'. More importantly, 84% pointed out that interoperability and information dynamic improvement should be of paramount importance in the relevant research and development. In addition, the more sustainable designs are becoming complicated, the more technical properties designers would employ. For example, they need to see the thermal or visual properties of building components and how they affect design sustainability. Presently, BIM is, to some extent, insufficient in consisting of properties to analyse energy, lighting and water efficiencies inherently. So, embedding this capability into a model design will become invaluable in the future (104).

However, this type of improvement can act as a double-edged sword. On the one hand, the fundamental tenet in the success of this end is having an integrated, comprehensive and organised system of collection and dissemination of information among the tools and parties involved (77). On the other hand, pushing to keep all information, either relevant or irrelevant, may be a source of friction and errors (105). The idealism concept about BIM, promoted by commercial sectors, delves into this misconception that all information is required and the more the merrier. They consist of detailed descriptions of all components that are often too 'information-rich' to evolve into the design, especially where continual changes happen (106). Likewise, observing the current trend of BIM platform development, designers notice drive-by software developers promote all-around tools and assist users in their procedure throughout the project lifecycle (107).

This idealistic introduction of BIM possibly leads to a quick uptake in AEC instead of a critical discourse concerning the adoption, mobilisation and simplification methods (108). However, according to the McGraw-Hill construction report of BIM contribution to green building, 54% of the participant companies claimed that the technical complexity of BIM is the reason behind not using it for green projects (109). Consequently, by overloading data or excessively technical sophistication in designing or exchanging BIM models, the receivers may have unreasonable expectations regarding the value of the information they have received (39).

Given the preceding context, in line with sustainable design leaning more and more toward intelligence, a significant breakthrough should be made in the current BIM to be smarter but simpler (110). Therefore, apropos of interdisciplinary application and the integration of sustainability, information and optimisation theories and their practical drivers of sustainable construction, BIM and AI, optimisation theory suggests a good convergence. It helps refine, optimise and classify the batch of information content in BIM through practical drivers such as AI and computational algorithms (111). Because of its parametric nature, BIM is a convenient platform for utilising AI in expert system developments. It can be based on the black box methods for predicting the current trend and optimising the future performance of design (112).

2.4 Artificial Intelligence (AI)

Rich and Knight (113) defined AI as 'the study of how to make computers do things at which people are doing better'. McCarthy (1968, p. 5) coined the term AI and gave a more primitive definition as 'the science and engineering of making intelligent machines'. In recent decades, several approaches, including linear, non-linear, heuristic and metaheuristic methods, have been developed based on AI as the alternatives for traditionally mathematical methods (114). Great trust in AI is the development of algorithmic models to assist with optimisation theory and its applications in refining data and automatically finding the most optimum choices (115).

In principle, AI applications in engineering problems can be categorised into three main areas (116):

- Prediction
- Classification
- Optimisation

Prediction and classification are two relatively similar methods of pre-processing data analysis used to describe important classes of data or to predict future data trends. However, while classification is usually used to classify categorical data, prediction models continuous-valued functions (114). Hence, these techniques need training data first to learn and, second, to infer the results based on inputs. ANN, DT and fuzzy systems are among the popular AI methods used for these purposes (117).

Furthermore, optimisation methods can be applied and considered a post-processing data step to achieve the best option among different choices. It is an iterative procedure that compares various solutions until an optimum or satisfactory resolution is found (28). Well-known used methods like GA and particle swarm optimisation (PSO) are inherent in mimicking biological systems and are appropriate for multivariable and non-linear problems (118).

2.4.1   AI Application in Sustainable Construction

Academics and professionals have employed AI in sustainable construction for the following aims:

- *Forecasting.* Different studies have been conducted to forecast electrical loads of utilities, short-term and long-term energy use, weather variables and thermal load for buildings (119–121). Typically, conditional demand analysis (CDA), ANN and support vector machine (SVM) are used for forecasting because of their strong prediction capability (122, 123).
- *System modelling.* ANN and fuzzy logic are often used for modelling various engineering systems in buildings such as HVAC, which can find an authentic relationship between inputs and outputs (124–127). ANN can be used for both continuous and discrete variables, but fuzzy logic is used for continuous ones (123).
- *Optimising.* Finding the most optimum values and states for different building design and operation problems is a real risk that AI can mitigate (128). Subsequently, evolutionary algorithms including GA and PSO are widely used for optimising building envelopes, massing, thermal comfort, daylighting and lifecycle analysis (129–132).

2.4.2   AI Application in BIM

Currently, most BIM features focus on the visualisation function and information exchange with other simulative and analytic software. Designers can easily identify the visual status of projects by the current BIM environment, but they cannot sufficiently obtain detailed scrutiny regarding the sustainability status of that project (76). Besides, the current BIM is regarded as passive as it requires an additional analysis procedure to assess and optimise its post-visualisation sustainability (133). Passive BIM cannot sufficiently provide sustainability analysis data such as energy consumption level and the design variables that need to be optimised. It has been proven that half of the time required for simulating and analysing a sustainability analysis model is spent re-creating BIM geometry via a new tool (73).

According to BuildingSmart (134): 'During the design period, different options are evaluated and tested. In a project using BIM, the model can be used to test "what if" scenarios and determine what the team will accomplish'. This is regarded as an 'active BIM' (135), which can guarantee the design's sustainability through facilitating 'what if' scenario tests that reflect on its design parameters. Hence, it can be expected that the active BIM would be successfully applied as a decision-making tool for GB. However, controllable parameters during the design phase, including physical properties and building envelop, have not been adequately managed under the present passive BIM. This fact leads to restrictions for the active application of BIM in critical processes such as daylighting and building energy simulation (136).

If these critics classify the current BIM as a passive system, the 'active BIM' system will provide diverse decision-making platforms to offer the optimal green design and construction (135). As a simple rule, the smarter the BIM, the more useful information it will contain specific to each of its contributors (110). That information needs to be managed, coordinated and associated with individual objects in the BIM to achieve a high level of usefulness (104). Leveraging the parametric definition inherited in the BIM-based sustainability analysis, AI, through optimal use of IT in preventing data duplication and malfunction, transforms data to an optimised level and contributes to the sustainable building (137). However, little concern has been given to applying AI in BIM to transform passive BIM into active BIM, especially sustainability analysis. In the literature, AI has been recently used in BIM for strengthening its parametric study for the following purposes:

- Optimising BIM-based construction planning and scheduling (135)
- Optimising BIM models based on different variables (111, 112)
- Optimising HVAC control schedule for building energy management systems (138)

These studies are in embryonic steps, but it is apparent that AI-based parametric analysis for active BIM function is a new trend that requires considerable progress to aid sustainable design. Nevertheless, this procedure can empower BIM with creative exploration of different design alternatives and automatically modify the variables and their interrelationships to make projects more sustainable.

2.5   Calculative, Simulative, Predictive and Optimisation Methods for Energy Efficient Buildings

As mentioned in Section 2.2, sustainable construction is broad, from green policies to renewable energy initiatives. However, figures such as 40% of total energy and 54% of electricity consumptions (139) and 9 Gt of carbon dioxide (CO_2) emissions as a harmful consequence (140) put energy efficient building design in the spotlight of research and development. Although energy consumption should not be treated as the single criterion for sustainability decision-making (141), within the majority of GB rating tools such as LEED, the most available points are given to the energy-oriented criteria (142). Furthermore, the potential of saving energy through appropriate measures in design and construction is perceived to be substantial and can reach up to 40% (143).

Concerning the International Energy Outlook of Energy Information Administration (EIA), the energy consumption of the building sector will increase by 48% in the next two decades at a ratio of 1.4% per annum (144). The building sector covers various building types such as residential, commercial and industrial. According to Lombard, Ortiz and Pout (145), the energy consumption from residential sectors for both developed and developing countries will be overtaking that of non-residential parts by 67% more in 2030.

Concerning these certain facts as evidence of significance, familiarity with EED and construction for residential buildings has been steadily increasing in recent years. Several methods and techniques in the scientific community have been developed, tested and verified for calculation, simulation, prediction and optimisation of buildings energy performance.

2.5.1   Calculative Methods

Calculative methods measure buildings' thermal behaviour through three main approaches of computational fluid dynamic (CFD), zone and multizone modes (146). CFD is a microscopic and 3D method to model thermal transfers and detailed flow fields. As an advantage, it is well performed for the precise analysis of the thermal performance of complex geometries. Still, it is very time consuming and requires deep knowledge of fluid dynamics and software (147). The zonal approach is a simplified version of CFD that divides each building zone into assorted cells corresponding to small parts of rooms in two dimensions. It can determine the local parameters in a 2D map and visualise airflows.

Nonetheless, it cannot deliver accurate results (148). The last and probably the simplest approach, as to its one-dimensional nature, is the multizone or nodal mode, which approximates a zone to a node specifying heating or cooling loads. The calculation time can be immensely decreased, linearising the equations. At the same time, the interrelationships of loads and their impacts are not quantifiable (149).

Looking briefly at the calculative methods, CFD is the complete approach to formulate building energy performance but, at the same time, is the most complex one. It hinders the simulation of all phenomena due to the extreme computation time (147). Zonal mode is an intermediate method that delivers less accurate results than the CFD but a more decent outcome than nodal mode (150). For all calculative approaches, a wide range of parameters such as geographical, physical, climatic and occupancy data are required as inputs. These variables are usually presented by a degree of uncertainty (151). Having stated that, detailed analysis of the physical characteristics of a target is the main shortcoming of these methods that necessitates extensive knowledge on the physical mechanisms of heat, air and moisture transfers (152).

2.5.2   Simulative Methods

A growing number of simulative methods have been developed and tested in the context of different simulation software in recent years to evaluate building energy performance. EnergyPlus, DOE-2.1E, Ecotect, EcoDesigner, Green Building Studio (GBS), Integrated Environmental Solution (IES), Modelica and DesignBuilder are the most popular software among the myriad tools that are majorly used by experts[2] (153). However, inconsistency in the outcomes and functions of these tools is one of the problems damaging their reliability. Raslan and Davies (154) revealed the predictive deviation for such software concerning the compliance check with energy efficiency standards and guidelines.

In addition, there is significant time required for learning and achieving proficiency in using these applications. By reviewing the procedural instruction of these applications, it can be inferred that designers should create the models in archi-tectural and construction tools and then follow the energy assessment by import-ing the models into these energy analysis applications (155). Thus, Schlueter and Thesseling (156) pointed out a need to seamlessly integrate this simulation practice into the design process. BIM may consist of almost the whole data required for an energy assessment. If it is employed properly, a great volume of time and tasks can be saved in preparation of the model for building energy simulation while minimising inaccuracies. This is seen as a necessity to enable the replacement of the traditional sequential processes with more interactive and concurrent design.

2.5.3 Predictive Methods

The peculiarity of predictive methods compared to calculative and simulative ones is that they do not necessitate gathering detailed physical information, working out heat transfer equations or simulating a complex energy model (127). Instead, they are based on machine learning algorithms, a subgroup of AI, deducing pat-terns only from training datasets to predict the energy consumption of buildings. Hence, they are well performed when the target building's comprehensive physi-cal, geometrical or geographical characteristics are not fully recognised (122). On the contrary, the predictive methods completely depend on the measures and reliability of data. If these data are neither accessible nor properly collected, it will be a serious concern (157). A wide range of algorithms is used for building performance prediction. However, according to a thorough review conducted by a group of prominent experts, three methods of CDA, ANN and SVM have been identified as the most popular and suitable methods in this regard (123).

CDA is a linear multivariable regression method that is used principally to fore-cast y as a linear aggregation of input parameters in addition to an error constant ε_i.

$$y_i = x_0 + x_1 \times x_{i1} + x_2 \times x_{i2} + \ldots + x_p \times x_{ip} + \varepsilon_i \quad i \in [1, n],\tag{2.1}$$

where n is the number of datasets, p is the number of parameters and x_0 is a bias (158).

Although CDA is a very simple method for predicting building energy perfor-mance by non-expert practitioners, it is very limited where the problem is non-linear (159). This drawback leads to inflexibility in prediction and trouble handling the multicollinearity issue, which is a correlation among variables. So, the lack of flexibility in its structure due to its linear nature makes it impractical to be utilised for the whole building energy performance (160).

The second method, ANN, is an AI-based computational model that tries to simulate biological neural networks' structure or functional aspects (161). The thermal equations used to analyse and calculate heat loads are often complex, making ANN a good platform for mitigating this complexity (162). In this form, the network is established with collected datasets, and the weights of inputs are fed into each neuron or nod. The consequences are then iteratively adjusted through feed forward-back propagation until a suitable output is produced. An appropriate production is the one that is as close as the actual one and benefits from the least error among the network outputs and target values (163). Back propagation is a method that feeds back the size of the error into the calculation for the weight changes.

Each neuron in the input layer represents one variable; *Input$_i$* multiplied by a weight, *W$_{ij}$*, and the results are added to give the output in the *Output$_j$* layer:

$$Output_j = f\left(\sum_i^j Input_i \times w_{ij} + B_j\right), \tag{2.2}$$

where f is the activation function that is converting weighted input to an output function and B_j is the bias for the *Output$_j$* (Figure 2.4).

ANN is greatly powerful in deducing the relationships between inputs and finding the relevant pattern toward the output through training procedure, without requiring a prior assumption or hypothesis and without being threatened by the multicollinearity phenomenon (164). Moreover, it minimises the discretisation error, which is caused by representing continuous variables in the computer with an integer value (117). Given this effective faculty, ANN can efficiently function a diversified range of variables from constant to binary and yes/no types. In contrast, requiring a huge and consistent database is a deficiency that limits ANN application. Therefore, it is essential to train the model with a comprehensive learning

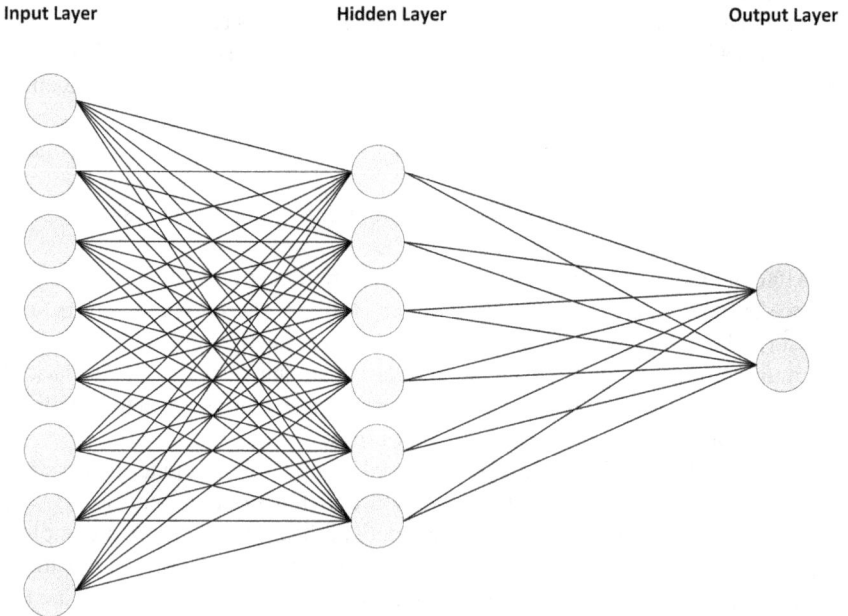

Figure 2.4 The Conceptual Structure of ANN

dataset, comparable data for all variables and no missing information (homogenous database) (165). The other barrier that may hinder its functionality lies in the lack of accredited rules in arranging the structure. Although some heuristic rules are suggested for selecting architecture for a neural network (166), some criteria such as the number of hidden layers and the associated neurons still come down to a rule of thumb and trial and error process.

SVM is another predictive and AI-based method used for prediction and forecasting purposes. In this regard, the assumption of SVM is to find the best fitting equation model for a training data of $[(x_1, y_1), (x_2, y_2), \ldots, (x_n, y_n)]$, in which x_i is in the input and y_i is in the output space (167). The function f for this dataset is as follows:

$$f(x) = \langle \omega, \varphi(x) \rangle + b,$$ (2.3)

where φ indicates a variable in the three-dimensional space, $< \ldots >$ a scalar function and ω and b are the constants for the linear combination of relative percentage error of the training data (168).

Estimating these constants is the main complexity in developing SVM as they correspond to the dots produced by scalar function in a three-dimensional space. So, the regularisation process is somewhat difficult for beginners (169). On the other hand, the flexibility in dealing with a heterogeneous database can alleviate its difficulty in establishing the process (167). This method can be run and trained through various data types with different natures. There is usually no restriction on the database except that vector data are required. As a huge advantage, it supports a database where all variables do not have the same amount of information or where we can find missing data (170).

As a holistic view of the predictive methods, multiple linear regression is probably easier. It can give good predictions and does not need real expertise to be implemented. But it is hugely limited because it assumes a linear description of the phenomenon (171). ANN overcomes the linearity problem. Nevertheless, it runs as a black box system, making interpretability very difficult (172).

Moreover, an important drawback of the ANN is that it requires a large amount and complete learning data (161). In contrast, SVM has the huge advantage of not necessitating complete data. However, contrary to ANN, it requires assuming the constants (123).

2.5.4   Optimisation Methods

Numerous methods are applied for optimising the energy consumption of buildings with a focus on different parameters. However, GA and PSO are very popular among the direct search, evolutionary methods and bio-inspired algorithms in sustainable building design (173).

GA, originally developed by Holland (174), uses adaptive heuristics to solve optimisation problems by mimicking the principles of natural selection. First, it starts with a randomly generated initial population of potential solutions for the optimisation problem. These possible solutions, often called chromosomes, are coded as binary or real strings. A new population of children chromosomes is then generated from the parent chromosomes through crossover and mutation procedures (175). While the crossover process combines parent chromosomes to produce children's chromosomes, the mutation consists of local chromosomes modifications. Next, the selection of the chromosomes is achieved based on their fitness values (Figure 2.5). Finally, the search of GA is terminated using a convergence threshold within a tolerable number of generations (176). Compared to conventional optimisation methods, GA has several distinctive features (177):

- Probabilistic rather than deterministic transition rules are applied to develop potential solutions from the current one.
- No derivatives are required; thus, any non-smooth objective function can be optimised.
- Global search can be used to avoid local optimum points.

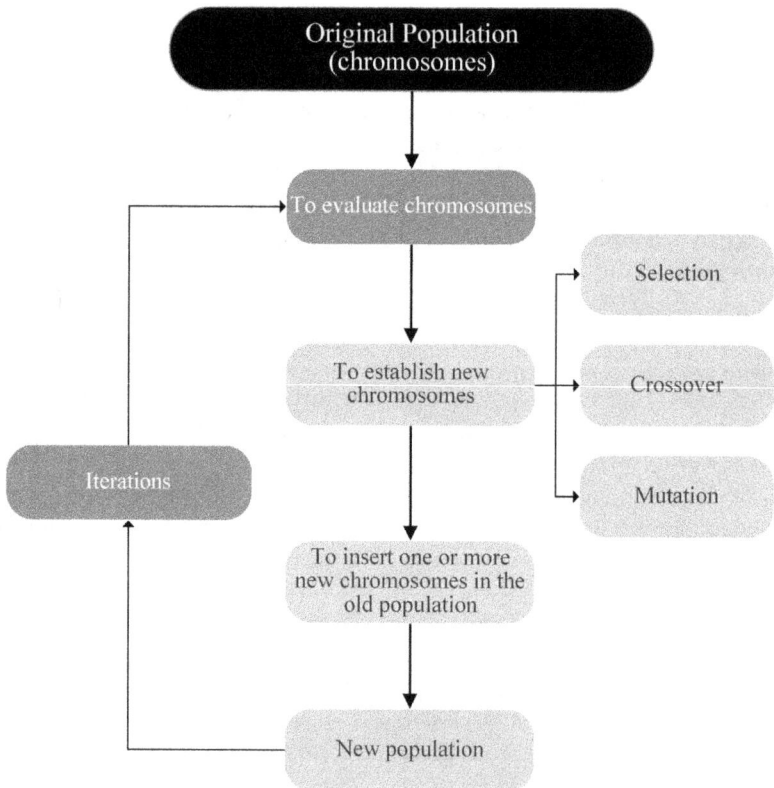

Figure 2.5 The Conceptual Procedure of GA

Source: Adapted from Foucquier et al. (123)

An important advantage of GA is that it deals with a powerful optimisation method to resolve energy problems and provides the convexity of the describing function (178). Another essential advantage is GA's ability to give several final solutions to an optimisation problem using a large number of input parameters. It allows the user to choose the most probable one (179). This can also be a drawback as the user can never be sure whether one has selected the best solution, especially as GA may not necessarily generate the optimal solution (175). GA needs considerable computational time, especially when practitioners address several variables ranging from building envelop to climatic data (116). It further suffers from the manual adjustment for the algorithm. Indeed, no rules exist to determine the number of individuals in the population, the number of generations or crossover and mutation probability. So, the only way to adjust the model is to test different combinations (31).

PSO, introduced by Kennedy and Eberhart (180), is a population-based stochastic optimisation technique. It works on the movement of swarms, inspired by the social behaviour of insects, animals herding, bird flocking and fish schooling, where a collaborative search for food has the potential for a computational model (181). Compared with other metaheuristic algorithms such as GA, PSO has less complicated operations and fewer defining parameters. Moreover, it can be coded in just a few lines and is highly dependent on stochastic processes (182). Because of its simplicity and performance, PSO has received increasing attention toward building energy optimisation (183).

In PSO, particles are the potential solutions of the optimisation problem, like points in the search space (initialisation). Compared to GA, PSO updates the particles iteratively by considering their internal velocity and the position obtained by the experience of all particles (184). Each particle's movement is influenced by its local best-known work but is also guided toward the most prominent positions in the search space. Particles are, then, updated for better positions and are found by other particles through constant evaluation and updates. This process is iteratively conducted to move the swarm toward the best solutions (185). PSO performs very well in global searches for a near-optimal solution for building design, especially when combined with other metaheuristic algorithms for simultaneous prediction and optimisation (186). However, increasing the number of particles prevents them from a quick cluster in the search, leading to more time for simulation and computation (187).

2.6   Summary

This chapter was set to introduce the book's three underlying theoretical domains, namely, sustainability, information and optimisation theories. These theories are mostly implemented sporadically in the built environment field via their practical drivers of sustainable construction, BIM and AI. However, the missed element here is their effective trilateral interaction toward an integrated manner to contribute to energy efficient building design. Henceforth, in line with the aim and objectives, a thorough understanding of the core concepts and sub-concepts of the EED is necessary to identify the current processes and methods. Furthermore, a more detailed and comprehensive analysis should support the current need for developing an active BIM function under AI to deliver more energy efficient and optimised residential buildings.

Notes

1 Creative Common License: CC BY-NC 4.0.
2 Fully detailed analysis of the simulative methods will be presented in Chapter 3.

References

1 Ding GKC, Banihashemi S. Ecological and carbon footprints: The future for city sustainability A2. In: Abraham MA, editor. Encyclopedia of sustainable technologies. Elsevier; 2017. p. 43–51.
2 Maczulak AE. Sustainability: Building eco-friendly communities: Infobase Publishing; 2010.
3 Pawlowski CW, Fath BD, Mayer AL, Cabezas H. Towards a sustainability index using information theory. Energy. 2005;30(8):1221–31.
4 Waseem N, Kota S, editors. Sustainability definitions: An analysis. International Conference on Research into Design, Springer; 2017.
5 Brundtland GH. World commission on environment and development: Our common future: Oxford University Press; 1987.
6 Kates R, Clark W. Our common journey: A transition toward sustainability: National Academy Press; 1999.
7 Glavič P, Lukman R. Review of sustainability terms and their definitions. Journal of Cleaner Production. 2007;15(18):1875–85.
8 Newman P. Can the magic of sustainability revive environmental professionalism? Greener Management International. 2005;2005(49):11–23.
9 Faucheux S, O'Connor M. Valuation for sustainable development: Methods and policy indicators: Edward Elgar Publishing Ltd; 1998.
10 Bell S, Morse S. Sustainability indicators: Measuring the immeasurable?: Earthscan; 2008.
11 White MA. Sustainability: I know it when I see it. Ecological Economics. 2013;86:213–7.
12 Wise R, Fazey I, Smith MS, Park S, Eakin H, Van Garderen EA, et al. Reconceptualising adaptation to climate change as part of pathways of change and response. Global Environmental Change. 2014;28:325–36.
13 Marinova D, McGrath N. Transdisciplinarity in teaching and learning sustainability. In: Rationality in an uncertain world. Edition Sigma; 2005. p. 275–85.
14 Nicolescu B, editor. The transdisciplinary evolution of learning. Symposium on Overcoming the Underdevelopment of Learning at the Annual Meeting of the American Educational Research Association, Montreal, Canada; 1999.
15 Nicolescu B. The transdisciplinary evolution of the university, condition for sustainable development. SCAN-9809055; 1998.
16 Todorov V. System sustainability and development sustainability: Modelling problems (Is E-conometrics of sustainable development possible?). Management and Sustainable Development. 2006;18(3–4):136–40.

17 Todorov V, Marinova D, editors. Information theory perspective on modeling sustainability. System Sciences (HICSS), 2010 43rd Hawaii International Conference on, IEEE; 2010.

18 Gray RM. Entropy and information theory: Springer; 2011.

19 Cover TM, Thomas JA. Elements of information theory: John Wiley & Sons; 2012.

20 Miller JG, Miller JL. Applications of living systems theory. Systems Practice. 1995;8(1):19–45.

21 Chase CH. Cognitive science: An introduction: MIT Press; 1995.

22 Morris RG, Tarassenko L, Kenward M. Cognitive systems-information processing meets brain science: Academic Press; 2005.

23 Curtis G, Cobham D. Business information systems: Analysis, design and practice: Pearson Education; 2008.

24 Liu K, Nakata K, Harty C. Pervasive informatics: Theory, practice and future directions. Intelligent Buildings International. 2010;2(1):5–19.

25 Sarker RA, Newton CS. Optimization modelling: A practical approach: CRC Press; 2007.

26 Sun W, Yuan Y-X. Optimization theory and methods: Nonlinear programming: Springer; 2006.

27 Deb K. Multi-objective optimisation using evolutionary algorithms: An introduction. In: Multi-objective evolutionary optimisation for product design and manufacturing. Springer; 2011. p. 3–34.

28 Fister Jr I, Yang X-S, Fister I, Brest J, Fister D. A brief review of nature-inspired algorithms for optimization. arXiv preprint arXiv:13074186. 2013.

29 Coello CAC, Lamont GB, Van Veldhuizen DA. Evolutionary algorithms for solving multi-objective problems: Springer; 2007.

30 Neri F, Cotta C. Memetic algorithms and memetic computing optimization: A literature review. Swarm and Evolutionary Computation. 2012;2:1–14.

31 Rao SS, Rao S. Engineering optimization: Theory and practice: John Wiley & Sons; 2009.

32 Diwekar UM. Introduction to applied optimization: Springer Nature; 2020.

33 Komiyama H, Takeuchi K, Shiroyama H, Mino T. Sustainability science: A multidisciplinary approach: United Nations University; 2011.

34 Haymaker JR, editor. Opportunities for AI to improve sustainable building design processes. 2011 AAAI Spring Symposium Series; 2011.

35 Clark WC, Crutzen PJ, Schellnhuber HJ. Science for global sustainability: Toward a new paradigm; 2005.

36 BS1192. 1192: 2007: Collaborative production of architectural, engineering and construction information, code of practice. ISBN: 978 0 580 58556 2. British Standards Institute (BSI); 2008.

37 Shafiq MT, Matthews J, Lockley S. A study of BIM collaboration requirements and available features in existing model collaboration systems. Journal of Information Technology in Construction (ITcon). 2013;18:148–61.

38 De Beaugrande R. Text, discourse, and process: Toward a multidisciplinary science of texts: Longman; 1980.

39 Guttman M. The information content of BIM: An information theory analysis of Building Information Model (BIM) content. Perkins and Will Research Journal. 2011;3(2):29–41.

40 Neustadt LW. Optimization: A theory of necessary conditions: Princeton University Press; 2015.

41   Bribián IZ, Capilla AV, Usón AA. Life cycle assessment of building materials: Comparative analysis of energy and environmental impacts and evaluation of the eco-efficiency improvement potential. Building and Environment. 2011;46(5):1133–40.

42   Wang T, Seo S, Liao P-C, Fang D. GHG emission reduction performance of state-of-the-art green buildings: Review of two case studies. Renewable and Sustainable Energy Reviews. 2016;56:484–93.

43   Ericson B, Hanrahan D, Kong V. The world's worst pollution problems: The top ten of the toxic twenty. Blacksmith Institute and Green Cross; 2008.

44   Venkatarama Reddy B, Jagadish K. Embodied energy of common and alternative building materials and technologies. Energy and Buildings. 2003;35(2):129–37.

45   EPA. Green building US: Environmental protection agency; 2016. Available from: https://archive.epa.gov/greenbuilding/web/html/.

46   Alwaer H, Clements-Croome DJ. Key Performance Indicators (KPIs) and priority setting in using the multi-attribute approach for assessing sustainable intelligent buildings. Building and Environment. 2010;45(4):799–807.

47   Bonini S, Swartz S. Profits with purpose: How organizing for sustainability can benefit the bottom line. McKinsey on Sustainability & Resource Productivity. 2014;2(1):1–15.

48   Anastaselos D, Oxizidis S, Manoudis A, Papadopoulos AM. Environmental performance of energy systems of residential buildings: Toward sustainable communities. Sustainable Cities and Society. 2016;20:96–108.

49   Kibert CJ. Sustainable construction: Green building design and delivery: Wiley; 2012.

50   Levine J. Three trees; 2008. Available from: www.flickr.com/photos/jeremylevinedesign/2815611090/in/photostream/.

51   UNEP. SBCI (Sustainable Building and Climate Initiative). Vision for sustainability on building and construction: United Nation Environment Program; 2014. Available from: www.unep.org/sbci/AboutSBCI/annual_reports.asp.

52   Häkkinen T, Belloni K. Barriers and drivers for sustainable building. Building Research & Information. 2011;39(3):239–55.

53   Banihashemi S, Shakouri M, Tahmasebi MM, Preece C. Managerial sustainability assessment tool for Iran's buildings. Proceedings of the ICE: Engineering Sustainability. 2014;167(1):12–23.

54   Larsson J, Jansson J, Olofsson T, Simonsson P, editors. Increased innovation through change in early design procedures. 19th IABSE Congress, Challenges in Design and Construction of an Innovative and Sustainable Built Environment; 2016.

55   Yang E. Diffusion of innovation in sustainable building practices in construction projects and the role of major stakeholders. 2016.

56   Stenberg A-C. The social construction of green building; 2006.

57   Foxon T, Pearson P. Overcoming barriers to innovation and diffusion of cleaner technologies: Some features of a sustainable innovation policy regime. Journal of Cleaner Production. 2008;16(1):S148–S161.

58   Banihashemi S, Hosseini MR, Golizadeh H, Sankaran S. Critical Success Factors (CSFs) for integration of sustainability into construction project management practices in developing countries. International Journal of Project Management. 2017;35(6):1103–19.

59   Reed B. Shifting from 'sustainability' to regeneration. Building Research & Information. 2007;35(6):674–80.

60   van Egmond E. Innovation, technology and knowledge transfer for sustainable construction. Construction Innovation and Process Improvement. 2012;95.

61 Chang R-D, Soebarto V, Zhao Z-Y, Zillante G. Facilitating the transition to sustainable construction: China's policies. Journal of Cleaner Production. 2016;131:534–44.

62 Chassiakos A, Sakellaropoulos S. A web-based system for managing construction information. Advances in Engineering Software. 2008;39(11):865–76.

63 Sodagar B, Fieldson R. Towards a sustainable construction practice. Construction Information Quarterly. 2008;10(3):101–8.

64 Hong T, Koo C, Kim J, Lee M, Jeong K. A review on sustainable construction management strategies for monitoring, diagnosing, and retrofitting the building's dynamic energy performance: Focused on the operation and maintenance phase. Applied Energy. 2015;155:671–707.

65 Passer A, Wall J, Kreiner H, Maydl P, Höfler K. Sustainable buildings, construction products and technologies: Linking research and construction practice. The International Journal of Life Cycle Assessment. 2015;20(1):1–8.

66 Gan X, Zuo J, Ye K, Skitmore M, Xiong B. Why sustainable construction? Why not? An owner's perspective. Habitat International. 2015;47:61–8.

67 Kramers A, Höjer M, Lövehagen N, Wangel J. Smart sustainable cities: Exploring ICT solutions for reduced energy use in cities. Environmental Modelling & Software. 2014;56:52–62.

68 Kibert CJ. Sustainable construction: Green building design and delivery: John Wiley & Sons; 2016.

69 Todorovic MS, Kim JT. Buildings energy sustainability and health research via interdisciplinarity and harmony. Energy and Buildings. 2012;47:12–18.

70 Zanni M-A, Soetanto R, Ruikar K. Defining the sustainable building design process: Methods for BIM execution planning in the UK. International Journal of Energy Sector Management. 2014;8(4):562–87.

71 Howell I, Batcheler B. Building information modeling two years later: Huge potential, some success and several limitations. The Laiserin Letter. 2005;22.

72 Azhar S, Carlton WA, Olsen D, Ahmad I. Building information modeling for sustainable design and LEED® rating analysis. Automation in Construction. 2011;20(2):217–24.

73 Krygiel E, Nies B. Green BIM: Successful sustainable design with building information modeling: John Wiley & Sons; 2008.

74 Teicholz P, Sacks R, Liston K. BIM handbook: A guide to building information modeling for owners, managers, designers, engineers, and contractors: Wiley; 2011.

75 Underwood J, Isikdag U, Global I. Handbook of research on building information modeling and construction informatics: concepts and technologies: Information Science Reference Hershey; 2010.

76 Wong K-D, Fan Q. Building Information Modelling (BIM) for sustainable building design. Facilities. 2013;31(3/4):138–57.

77 Hardin B. BIM and construction management: Proven tools, methods, and workflows: John Wiley & Sons; 2011.

78 Woo J-H, Menassa C. Virtual retrofit model for aging commercial buildings in a smart grid environment. Energy and Buildings. 2014;80(0):424–35.

79 Hiyama K, Kato S, Kubota M, Zhang J. A new method for reusing building information models of past projects to optimize the default configuration for performance simulations. Energy and Buildings. 2014;73:83–91.

80 Park J, Park J, Kim J, Kim J. Building information modelling based energy performance assessment system: An assessment of the energy performance index in Korea. Construction Innovation: Information, Process, Management. 2012;12(3):335–54.

81 Seongchan K, Jeong-Han W, editors. Analysis of the differences in energy simulation results between Building Information Modeling (BIM)-based simulation method and the detailed simulation method. Simulation Conference (WSC), Proceedings of the 2011 Winter; December 11–14, 2011.

82 Huang YC, Lam KP, Dobbs G, editors. A scalable lighting simulation tool for integrated building design. Third National Conference of IBPSA-USA, Berkeley, CA; July 30–August 1, 2008.

83 Raminhos F, Valdez MT, Ferreira CM, Barbosa FM, editors. Energy efficiency in artificial light design using different computational simulation tools. Proceedings of International Conference on Engineering and Computer Education; 2013.

84 Welle B, Rogers Z, Fischer M. BIM-centric daylight profiler for simulation (BDP-4SIM): A methodology for automated product model decomposition and recomposition for climate-based daylighting simulation. Building and Environment. 2012;58(0):114–34.

85 Yan W, Clayton M, Haberl J, Jeong W, Kim JB, Kota S, et al., editors. Interfacing BIM with building thermal and daylighting modeling. Building Simulation Conference; 2013.

86 Tabadkani A, Banihashemi S, Hosseini MR, editors. Daylighting and visual comfort of oriental sun responsive skins: A parametric analysis. In: Building simulation. Springer: 2018.

87 Banihashemi S, Tabadkani A, Hosseini MR. Modular coordination-based generative algorithm to optimize construction waste. Procedia Engineering. 2017;180:631–9.

88 Banihashemi S, Tabadkani A, Hosseini MR. Integration of parametric design into modular coordination: A construction waste reduction workflow. Automation in Construction. 2018;88:1–12.

89 Wong KD, Fan Q. Building Information Modelling (BIM) for sustainable building design. Facilities. 2013;31(3/4):138–57.

90 Oh S, Kim Y, Park C, Kim I, editors. Process-driven BIM-based optimal design using integration of EnergyPlus, genetic algorithm, and pareto optimality. 12th Conference of International Building Performance Simulation Association, Sydney; 2011.

91 Wang J, Li J, Chen X, editors. Parametric design based on building information modeling for sustainable buildings. Challenges in Environmental Science and Computer Engineering (CESCE), 2010 International Conference on, IEEE; 2010.

92 Wu W, Issa R. BIM execution planning in green building projects: LEED as a use case. Journal of Management in Engineering. 2014;31:401–07.

93 Wu W, Issa R. Leveraging cloud-BIM for LEED automation. Journal of Information Technology in Construction. 2012;17:367–84.

94 Motawa I, Carter K. Sustainable BIM-based evaluation of buildings. Procedia: Social and Behavioral Sciences. 2013;74(0):419–28.

95 Nguyen T, Shehab T, Gao Z. Evaluating sustainability of architectural designs using building information modeling. The Open Construction and Building Technology Journal. 2010;4(1):1–8.

96 Dong B, Lam K, Huang Y, Dobbs G, editors. A comparative study of the IFC and gbXML informational infrastructures for data exchange in computational design support environments. Tenth International IBPSA Conference; 2007.

97 Inyim P, Rivera J, Zhu YM. Integration of building information modeling and economic and environmental impact analysis to support sustainable building design. Journal of Management in Engineering. 2015;31(1).

98 Clevenger CM, Khan R. Impact of BIM-enabled design-to-fabrication on building delivery. Practice Periodical on Structural Design and Construction. 2013;19(1):122–8.

99 Biswas T, Tsung-Hsien Wang RK. Integrating sustainable building rating systems with building information models. 2008.

100 Gandhi S, Jupp J. BIM and Australian green star building certification. Computing in Civil and Building Engineering. 2014:275–82.

101 Azhar S. BIM for sustainable design: Results of an industry survey. Journal of Building Information Modeling. 2010;4(1):27–8.

102 Malkin R. Sustainable design analysis and BIM [Building Information Modelling]. Architecture Australia. 2011;100(4):104.

103 Bynum P, Issa R, Olbina S. Building information modeling in support of sustainable design and construction. Journal of Construction Engineering and Management. 2013;139(1):24–34.

104 Kensek K, Noble D. Building information modeling: BIM in current and future practice: John Wiley & Sons; 2014.

105 Sanguinetti P, Abdelmohsen S, Lee J, Lee J, Sheward H, Eastman C. General system architecture for BIM: An integrated approach for design and analysis. Advanced Engineering Informatics. 2012;26(2):317–33.

106 Holzer D. BIM's seven deadly sins. International Journal of Architectural Computing. 2011;9(4):463–80.

107 Zeng MN. Future of green BIM designing and tools. Advanced Materials Research. 2012;374:2557–61.

108 Holzer D. Optioneering in collaborative design practice. International Journal of Architectural Computing. 2010;8(2):165–82.

109 Bernstein H, Jones S, Russo M. Green BIM: How building information modeling is contributing to green design and construction. McGraw-Hill Construction; 2010.

110 Heidari M, Allameh E, de Vries B, Timmermans H, Jessurun J, Mozaffar F. Smart-BIM virtual prototype implementation. Automation in Construction. 2014;39:134–44.

111 Asl MR, Bergin M, Menter A, Yan W. BIM-based parametric building energy performance multi-objective optimization. Fusion, Proceedings of the 32nd International Conference on Education and research in Computer Aided Architectural Design in Europe, Northumbria University, Newcastle upon Tyne, UK; September 2014. p. 455–64.

112 Yuan Y, Yuan J, Fan X. Integration of BIM and intelligence algorithm for BLC energy consumption evaluation and optimization: Principles and framework. Journal of Convergence Information Technology. 2013;8(10):502–9.

113 Rich E, Knight K. Artificial intelligence: McGraw-Hill; 1991.

114 Cohen PR, Feigenbaum EA. The handbook of artificial intelligence: Butterworth-Heinemann; 2014.

115 Engelbrecht AP. Introduction to computational intelligence: Computational intelligence: John Wiley & Sons, Ltd; 2007. p. 1–13.

116 Engelbrecht AP. Computational intelligence: An introduction: John Wiley & Sons; 2007.

117 Kantardzic M. Data mining: Concepts, models, methods, and algorithms: John Wiley & Sons; 2011.

118 Tao F, Zhang L, Laili Y. Configurable intelligent optimization algorithm: Springer; 2014.

119 Yang J, Rivard H, Zmeureanu R, editors. Building energy prediction with adaptive artificial neural networks. Ninth International IBPSA Conference, Montréal, Canada, August; 2005.

120 Kreider J, Claridge D, Curtiss P, Dodier R, Haberl J, Krarti M. Building energy use prediction and system identification using recurrent neural networks. Journal of Solar Energy Engineering. 1995;117(3):161–6.

121 Virote J, Neves-Silva R. Stochastic models for building energy prediction based on occupant behavior assessment. Energy and Buildings. 2012;53:183–93.

122 Zhao H-X, Magoulès F. A review on the prediction of building energy consumption. Renewable and Sustainable Energy Reviews. 2012;16(6):3586–92.

123 Foucquier A, Robert S, Suard F, Stéphan L, Jay A. State of the art in building modelling and energy performances prediction: A review. Renewable and Sustainable Energy Reviews. 2013;23:272–88.

124 Kalogirou SA. Applications of artificial neural-networks for energy systems. Applied Energy. 2000;67(1):17–35.

125 Kalogirou SA. Applications of artificial neural networks in energy systems. Energy Conversion and Management. 1999;40(10):1073–87.

126 Kalogirou SA. Artificial neural networks in renewable energy systems applications: A review. Renewable and Sustainable Energy Reviews. 2001;5(4):373–401.

127 Li K, Su H, Chu J. Forecasting building energy consumption using neural networks and hybrid neuro-fuzzy system: A comparative study. Energy and Buildings. 2011;43(10):2893–9.

128 Griego D, Krarti M, Hernández-Guerrero A. Optimization of energy efficiency and thermal comfort measures for residential buildings in Salamanca, Mexico. Energy and Buildings. 2012;54(0):540–9.

129 Machairas V, Tsangrassoulis A, Axarli K. Algorithms for optimization of building design: A review. Renewable and Sustainable Energy Reviews. 2014;31:101–12.

130 Palonen M, Hasan A, Siren K, editors. A genetic algorithm for optimization of building envelope and HVAC system parameters. Proc of the 11th IBPSA Conference, Glasgow, Scotland; 2009.

131 Evins R, Pointer P, Burgess S. Multi-objective optimisation of a modular building for different climate types. Proceedings of the Building Simulation and Optimisation; 2012.

132 Caldas L. Generation of energy efficient architecture solutions applying GENE_ARCH: An evolution-based generative design system. Advanced Engineering Informatics. 2008;22(1):59–70.

133 Moon HS, Kim HS, Kang LS, Kim CH. BIM functions for optimized construction management in civil engineering. Gerontechnology. 2012;11(2):67.

134 BuildingSmart. Integrated project delivery: A working definition: AIA California Council; 2007.

135 Moon H, Kim H, Kamat V, Kang L. BIM-based construction scheduling method using optimization theory for reducing activity overlaps. Journal of Computing in Civil Engineering. 2013:04014048.

136 Rahmani M, Zarrinmehr S, Yan W. Towards BIM-based parametric building energy performance optimization. Proceedings of the 33rd Annual Conference of the Association for Computer Aided Design in Architecture (ACADIA), ACADIA 13: Adaptive Architecture, Cambridge; 2013. p. 101–8.

137 Succar B, Sher W, Williams A. Measuring BIM performance: Five metrics. Architectural Engineering and Design Management. 2012;8(2):120–42.

138 Jung DK, Lee DH, Shin JH, Song BH, Park SH. Optimization of energy consumption using BIM-based building energy performance analysis. Applied Mechanics and Materials. 2013;281:649–52.

139 IEA. World energy outlook. Paris, France: International Energy Agency (IEA); 2012.

140 Jennings M, Hirst N, Gambhir A. Reduction of carbon dioxide emissions in the global building sector to 2050. Grantham Institute for Climate Change Report GR. 2011;3.

141 Ding GK. Sustainable construction: The role of environmental assessment tools. Journal of Environmental Management. 2008;86(3):451–64.

142 Pohl J. Sustainable architecture concepts and principles. In: Building science. Blackwell Publishing Ltd.; 2011. p. 225–58.

143 Costa A, Keane MM, Torrens JI, Corry E. Building operation and energy performance: Monitoring, analysis and optimisation toolkit. Applied Energy. 2013;101:310–16.

144 EIA. International energy Outlook energy information administration; 2006.

145 Pérez-Lombard L, Ortiz J, Pout C. A review on buildings energy consumption information. Energy and Buildings. 2008;40(3):394–8.

146 Pedersen L. Use of different methodologies for thermal load and energy estimations in buildings including meteorological and sociological input parameters. Renewable and Sustainable Energy Reviews. 2007;11(5):998–1007.

147 Zhai ZJ, Chen QY. Performance of coupled building energy and CFD simulations. Energy and Buildings. 2005;37(4):333–44.

148 Inard C, Meslem A, Depecker P. Energy consumption and thermal comfort in dwelling-cells: A zonal-model approach. Building and Environment. 1998;33(5):279–91.

149 Megri AC, Haghighat F. Zonal modeling for simulating indoor environment of buildings: Review, recent developments, and applications. Hvac&R Research. 2007;13(6):887–905.

150 Zhai Z, Chen Q, Haves P, Klems JH. On approaches to couple energy simulation and computational fluid dynamics programs. Building and Environment. 2002;37(8):857–64.

151 Swan LG, Ugursal VI. Modeling of end-use energy consumption in the residential sector: A review of modeling techniques. Renewable and Sustainable Energy Reviews. 2009;13(8):1819–35.

152 Woloszyn M, Rode C, editors. Tools for performance simulation of heat, air and moisture conditions of whole buildings. In: Building simulation. Springer; 2008.

153 EERE. Building energy software tools directory: U.S. Department of Energy; 2014. Available from: http://apps1.eere.energy.gov/buildings/tools_directory/.

154 Raslan R, Davies M. Results variability in accredited building energy performance compliance demonstration software in the UK: An inter-model comparative study. Journal of Building Performance Simulation. 2010;3(1):63–85.

155 Kavgic M, Mavrogianni A, Mumovic D, Summerfield A, Stevanovic Z, Djurovic-Petrovic M. A review of bottom-up building stock models for energy consumption in the residential sector. Building and Environment. 2010;45(7):1683–97.

156 Schlueter A, Thesseling F. Building information model based energy/exergy performance assessment in early design stages. Automation in Construction. 2009;18(2):153–63.

157 Neto AH, Fiorelli FAS. Comparison between detailed model simulation and artificial neural network for forecasting building energy consumption. Energy and Buildings. 2008;40(12):2169–76.

158 Aigner DJ, Sorooshian C, Kerwin P. Conditional demand analysis for estimating residential end-use load profiles. The Energy Journal. 1984:81–97.

159 Ghiaus C. Experimental estimation of building energy performance by robust regression. Energy and Buildings. 2006;38(6):582–7.

160 Aydinalp-Koksal M, Ugursal VI. Comparison of neural network, conditional demand analysis, and engineering approaches for modeling end-use energy consumption in the residential sector. Applied Energy. 2008;85(4):271–96.

161 Wang S-C. Artificial neural network. In: Interdisciplinary computing in java programming. Springer; 2003. p. 81–100.

162 Ahmad A, Hassan M, Abdullah M, Rahman H, Hussin F, Abdullah H, et al. A review on applications of ANN and SVM for building electrical energy consumption forecasting. Renewable and Sustainable Energy Reviews. 2014;33:102–9.

163 Hagan MT, Demuth HB, Beale MH. Neural network design: Pws Pub.; 1996.

164 Dreiseitl S, Ohno-Machado L. Logistic regression and artificial neural network classification models: A methodology review. Journal of Biomedical Informatics. 2002;35(5):352–9.

165 Dounis AI. Artificial intelligence for energy conservation in buildings. Advances in Building Energy Research. 2010;4(1):267–99.

166 Shahidehpour M, Yamin H, Li Z. Market operation in electric power systems: Wiley-IEEE Press; 2002.

167 Smola AJ, Schölkopf B. A tutorial on support vector regression. Statistics and Computing. 2004;14(3):199–222.

168 James G, Witten D, Hastie T, Tibshirani R. An introduction to statistical learning: Springer; 2013.

169 Meyer D, Leisch F, Hornik K. The support vector machine under test. Neurocomputing. 2003;55(1):169–86.

170 Li Q, Meng Q, Cai J, Yoshino H, Mochida A. Applying support vector machine to predict hourly cooling load in the building. Applied Energy. 2009;86(10):2249–56.

171 Clarke J. Energy simulation in building design: Routledge; 2007.

172 Tso GK, Yau KK. Predicting electricity energy consumption: A comparison of regression analysis, decision tree and neural networks. Energy. 2007;32(9):1761–8.

173 Evins R. A review of computational optimisation methods applied to sustainable building design. Renewable and Sustainable Energy Reviews. 2013;22:230–45.

174 Holland JH. Adaptation in natural and artificial systems: An introductory analysis with applications to biology, control, and artificial intelligence: University Michigan Press; 1975.

175 Davis L. Handbook of genetic algorithms: Van Nostrand Reinhold New York; 1991.

176 Gendreau M, Potvin J-Y. Handbook of metaheuristics: Springer; 2010.

177 Sumathi S, Paneerselvam S. Computational intelligence paradigms: Theory & applications using MATLAB: CRC Press; 2010.

178 Houck CR, Joines J, Kay MG. A genetic algorithm for function optimization: A Matlab implementation. NCSU-IE TR. 1995;95(09).

179 Caldas LG, Norford LK. Genetic algorithms for optimization of building envelopes and the design and control of HVAC systems. Journal of Solar Energy Engineering. 2003;125(3):343–51.

180 Kennedy J, Eberhart R, editors. Particle swarm optimization. Proceedings of IEEE International Conference on Neural Networks, Perth, Australia; 1995.

181 Eberhart RC, Kennedy J, editors. A new optimizer using particle swarm theory. Proceedings of the Sixth International Symposium on Micro Machine and Human Science, New York, NY; 1995.

182 Marler RT, Arora JS. Survey of multi-objective optimization methods for engineering. Structural and Multidisciplinary Optimization. 2004;26(6):369–95.

183 Banos R, Manzano-Agugliaro F, Montoya F, Gil C, Alcayde A, Gómez J. Optimization methods applied to renewable and sustainable energy: A review. Renewable and Sustainable Energy Reviews. 2011;15(4):1753–66.

184 Shi Y, Eberhart RC, editors. Empirical study of particle swarm optimization. Evolutionary Computation, 1999 CEC 99 Proceedings of the 1999 Congress on, IEEE; 1999.

185 Kennedy J, Kennedy JF, Eberhart RC. Swarm intelligence: Morgan Kaufmann; 2001.

186 Bukhari F, Frazer JH, Drogemuller R, editors. Evolutionary algorithms for sustainable Building design. The 2nd International Conference on Sustainable Architecture and Urban Development, Amman, Jordan; 2010.

187 Wetter M, Wright J. A comparison of deterministic and probabilistic optimization algorithms for nonsmooth simulation-based optimization. Building and Environment. 2004;39(8):989–99.

3 BIM and Energy Efficient Design

3.1 Background

The building energy efficiency concept refers to the energy required to provide suitable environmental conditions (comfort band), which can minimise the energy consumption of a building facility (1). Hence, EED scrutinises the design aspects of such a concept and includes identifying design parameters, their optimisation and their application in the design process (2). To effectively put EED into action, it is always recommended to target the design stage among the building lifecycles. This stage is a very effective period of integrating energy-saving criteria into the design to minimise energy consumption and financial costs throughout the project life cycle (3).

In essence, the EED of a building facility should be a multi-objective and multivariable design effort. At the same time, the traditional trial and error design techniques rely on designers' knowledge and skills, which can be inefficient in complex design tasks (4). In terms of the architectural design methodology, two major elements of function and form consolidate the conventional design principles. The driving force behind this methodology is the architect's rationality and sensibility (5). However, EED of a building is a performance-driven process. It means that the driving force behind this design method is a quantitative performance figure such as the amount of energy consumption reduction (6). Therefore, this design methodology requires a more systematic and organised way of design generation to overcome the drawbacks of the conventional design. It should pave the way for a rapid and accurate energy performance calculation and a systematic technique for searching the optimal solution among a large design space (7).

Hence, EED mostly deals with two main elements of AI application in the design parameters optimisation and a design integration platform: CDE (4). AI-based optimisation alleviates the trial-and-error design routine by generating various designs and finding the optimal or near-optimal design solutions (8). On the other hand, the design integration platform built-in CDE establishes an automated workflow to apply different energy simulation and analysis packages to realise the optimisation outcomes in the design procedure (9). Therefore, more integrated energy-related design optimisation is imperative to effectively fit into the architectural design workflow and deliver the EED (10).

DOI: 10.1201/9781003207658-3

Thus, considering the above remarks, a BIM-enabled EED refers to a procedure involving relevant BIM processes and tools to generate, exchange and manage an energy efficient built environment. Success in delivering this concept largely relies on how effectively BIM adopts EED. As a result, BIM and EED have become a growing field of research, while a comprehensive analysis on the topic is still missing. In recent years, scholars have attempted to research BIM and EED (hereafter referred to as BIM-EED) potentials from a different point of view. As a result, this topic has been given considerable attention in the body of knowledge, which has resulted in a rapid rise in the number of related publications (11). Such a substantial increase in research on the topic presents a potential risk of failing to assess the status of the body of knowledge accurately. This becomes a major barrier to identifying required directions for research on the topic, thus increasing the potential for overlooking pivotal aspects, as Yalcinkaya and Singh (2015) asserted. In essence, the emerging body of existing research on BIM-EED warrants conducting a rigorous critical review to uncover certain research requirements, which have not been adequately met in the current research trends (12).

Furthermore, the body of the knowledge on BIM-EED is fragmented, with limited systematic attempts to evaluate it from a widely held view (13). Some reviews, such as broad studies of Foucquier, Robert (14) and Negendahl (15), have been for the most part qualitative, based on manual reviews. Thus, the findings are fairly prone to subjectivity (16). On top of that, some review studies have also focused one-sidedly on studies targeting narrowed aspects associated with the topic. For example, Shi, Tian (4) and Wang and Srinivasan (17) were restricted to studies on building energy modelling and prediction. Therefore, there is a conspicuous lack of quantified, methodical and comprehensive assessment of the body of the knowledge on BIM-EED. The present section aims to go beyond such attempts based on the state of the art, systematic, content, thematic and structured gap analyses to present the concise and precise picture of the research on BIM-EED. The findings facilitate the development and improvement of research on BIM for EED through discussing the current state of the art, existing gaps and themes and links between the concepts involved and the future research agenda.

3.2   The Current State of the Art of BIM-EED

According to the major stages of knowledge evolution around emerging technologies and their adoption cycles introduced by Day, Schoemaker and Gunther (18), the current state of the art of BIM-EED can be categorised into three main levels of BIM-compatible, BIM-integrated and BIM-inherited EED in terms of the BIM adoption processes. This is illustrated in Figure 3.1 and elaborated in the following subsections. This classification aligns the three phases of discovering, probing and learning, committing and competing (knowledge evolution steps) with the BIM-EED cycles. Furthermore, it is worth mentioning that this classification aligns with the data evolution stages toward CDE as the single source of BIM data. Therefore, BIM-inherited EED signifies CDE elaboration, and the ultimate aim is to develop the integration framework of active BIM with AI. The next sections fully discussed

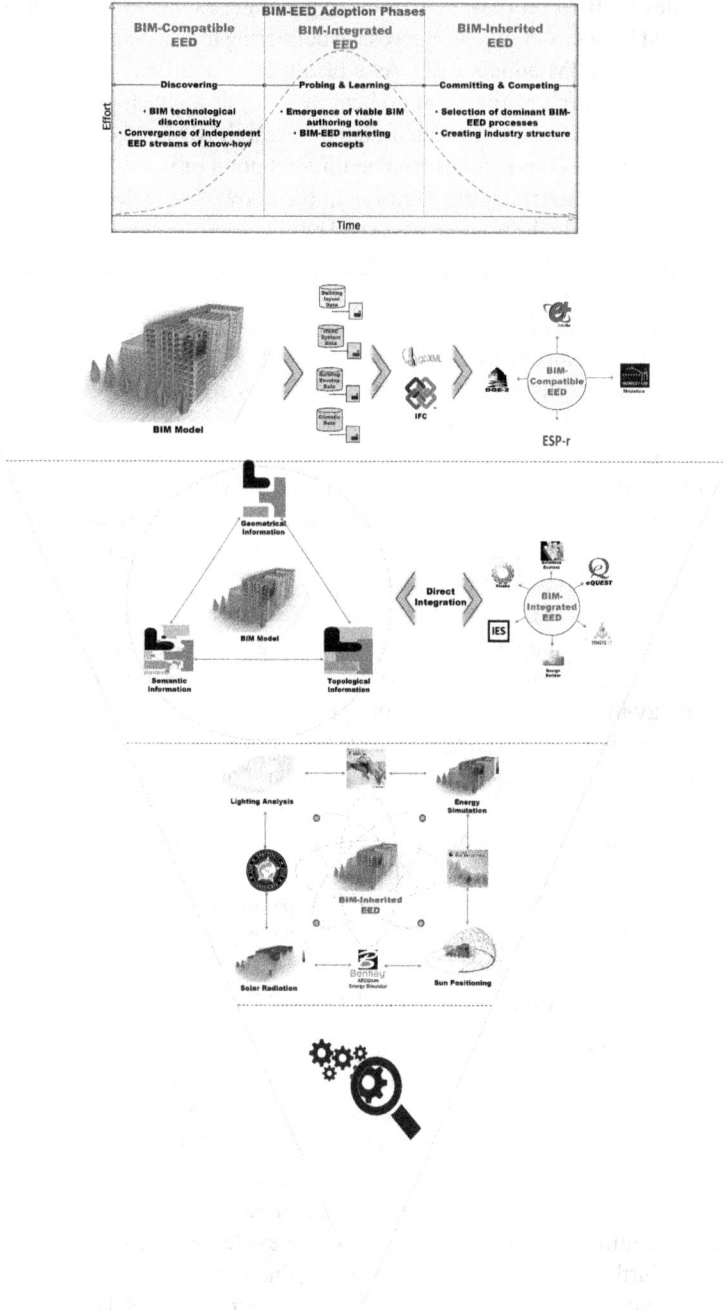

Figure 3.1 The Current State of the Art of BIM-EED
Source: Adapted from (18–20)

this evolution, the differences, contributions and the recommended approaches. Moreover, the simulation software, BIM adoption appreciation, interoperability methods used by each study and level of detail (LoD) of BIM models are identified.

3.2.1   BIM-Compatible EED

BIM-compatible EED is the first level of BIM adoption into EED, which commenced with Chaisuparasmikul (2006) and Zeng and Zhao (2006) by introducing the data exchanges possibility between building 3D models and EnergyPlus® application. This process can be generally summarised in BIM modelling of the case, extracting the available and relevant data of the model such as building layout, HVAC system, building envelope and climatic data and populating these data through data exchange platforms into building energy simulation tools (see Figure 3.1).

BIM-compatible EED involves the first generation of building energy simulation software, including EnergyPlus®, DOE2®, ESP-r® and Modelica®, which are considered as 'not BIM aware applications' due to the lack of an intimate relationship with a BIM authoring tool (19). This observation is corroborated by the BIM-EED adoption phases, which establish the foundation of this analysis and are adapted from a seminal study on managing emerging technologies (18). Figure 3.1 indicates that BIM-compatible EED corresponds to the discovering phase. It is featured with the BIM technological discontinuity and convergence of independent EED streams of know-how (21). The identified discontinuity necessitates a common file format as a hub of data exchange to the light in which one software writes and the other reads (21). Such a data transfer process may occur inadvertently or deliberately. For example, it may involve a non-proprietary data exchange platform such as Industry Foundation Class (IFC) or the registered energy data transfer platform of gbXML. This type of information exchange is not as rigorous as BIM aware tools because it is not specially designated and is unlikely to have been extensively tested (22).

3.2.2   BIM-Integrated EED

The second level of adopting BIM into EED can be labelled as BIM-integrated EED, mainly characterised by the direct integration between BIM authoring applications and energy simulation software (23). In this stage, intelligence maturity grows from data to the information level. It allows for understanding relations, dashboard visualisation and interpretation between the initial inputs of the BIM model and the results obtained from the energy analysis applications (24). Specifically, three groups – geometrical, topological and semantic – that contain the information required for an energy analysis assessment are integrated with the second generation of building performance simulation tools: IES®, DesignBuilder®, Trnsys 17®, eQuest®, Autodesk Ecotect® and Riuska® (17) (see Figure 3.1). These applications are 'BIM aware' as to the capability of making API (Application Programming Interface) calls to BIM authoring tools and extracting the information directly from the BIM model (21). IES® and DesignBuilder® are even among the

BIM-EED aware packages that update the data held in the BIM database and create a return trip of information cycle (25). Hence, it can be seen in Figure 3.1 that the emergence of viable BIM authoring tools leads to the probing and learning of BIM-EED concepts and shifts the efforts to the second phase of BIM-integrated EED.

3.2.3 BIM-Inherited EED

The current trend in applying BIM-EED processes and creating an industry-based structure for an energy efficient built environment result in committing and competing for the stage of BIM-EED adoption phases, which, in turn, embarks on BIM-inherited EED (Figure 3.1). This level of implementation focuses on delivering a strategic value of coherent and structured digital information of a project together with a package of building performance simulation (21). What distinguishes this process from the previous ones is the ability to perform such a package in the homogenous environment of BIM. Therefore, BIM market leaders develop a new generation of EED applications hinging on the utility of the underlying database of the BIM design repository. This category of BIM, which is adopted into EED, is internally interoperable through utilising the inbuilt information and includes Autodesk GBS®, Autodesk Formit 360®, Graphisoft EcoDesigner® and Bentley AECOsim Energy Simulator® (20). It is depicted in Figure 3.1, and there are four groups presented by BIM-inherited EED: energy simulation, solar radiation, sun positioning and lighting analyses.

3.2.4 BIM-EED Adoption

As mentioned earlier, there are three levels – BIM-compatible, BIM-integrated and BIM-inherited EED – that elaborate the BIM adoption processes into the energy efficient built environment. Given the importance of grasping clear perception toward the current status of each group of BIM-EED, the analysis shows that the majority of the existing application falls to the BIM-compatible group, while BIM-integrated and BIM-inherited EEDs are in the next priorities. This fact aligns with the knowledge evolution of BIM as an emerging technology and its adoption cycles (18). Holistically, the key of the shift from BIM-compatible to the BIM-integrated and BIM-inherited EEDs is in the light of the growing BIM implementation levels in AEC, from level 1 to level 3, and it is expected that the current state will follow the same trend in the future (11).

Technically, BIM adoption into the EED procedure is tied to the advent of the different energy simulation software and their performance rate and effectiveness in the application phase. As illustrated in Figure 3.1, each level of BIM-EED is run through the specific software or procedure, and hence, the role of simulation is at the heart of our understanding of BIM-EED adoption. Building performance simulation can be defined as 'the science of estimating future states of single or multiple physical phenomena within an existing or proposed built environment' (26). Considering this definition, the 'tool function wish list' characterises a comprehensive, efficient and precise building performance simulation as the following (27):

- Supporting design simulation decision-making under risk and uncertainty
- Recognising regular analysis with minimum variations and supporting incremental design strategies
- Embedding application validity rules to detect whether the application results beyond its validity range
- Performing robust solvers for non-linear, mixed and hybrid simulations

The technical criteria for assessing these characteristics can then be framed into the accuracy and validity of the simulation, transferability (interoperability) of the model and data and confirmability of the construction and information level behind the simulation model (26).

3.2.5 *Simulation Software*

As mentioned in the previous sections, various simulation software applications are employed in BIM-EED adoption levels for building performance simulation purposes. EnergyPlus®, DOE 2® and Modelica® are the most popular software among the tools categorised as the first group, BIM-compatible EED. EnergyPlus® is a simulation engine imputing and outputting text files to calculate heating and cooling loads in a variable point of time. It provides accurate results, allowing users to evaluate the radiant heating, cooling and inter-zonal loads. However, it hugely suffers from the lack of a graphical user interface (28). Nevertheless, EnergyPlus® maintains its first place in BIM-compatible EED as to its great accuracy.

DOE 2® is the second rank of this software category, which forecasts hourly energy consumption and associated costs based on the climatic data, HVAC information and building geometry. It benefits from a precise building energy simulation engine, but employing four different simulation sub-programs (Loads®, Systems®, Plant® and Econ®) makes it a very complicated tool for application (19). Modelica® is the next priority and is an object library based, free, open-source dynamic simulation model for building energy and control systems. It is fast and flexible but does not present an interface to the user.

The second group, BIM-integrated EED, is mostly addressed in the literature using Ecotect®, IES® and DesignBuilder®. As a highly visualised tool, Ecotect® benefits from a powerful 3D modelling engine. It covers the thermal, daylighting, acoustic and cost analysis and allows for real-time animation features, though it is argued that Ecotect® cannot rigorously estimate the heating and cooling loads of buildings and in the incremental design steps due to the weaknesses in its thermal engine (29). This fact warns the professional and scientific community about the reliability of the simulation results with Ecotect®.

IES® is another tool that is reputed to be one of the most 'architect friendly' building performance simulation tools based on Attia, Beltrán (30) and is further recognised as the strongest tool based on its range of analysis options (31). Nevertheless, it suffers from a lack of interoperability with the IFC exchange protocol (32). Finally, DesignBuilder® is another energy simulation package within BIM-integrated EED. It is a powerful parametric building energy modelling tool that exploits a direct integration with BIM authoring tools and the rapid interoperability with non-BIM aware applications through its gbXML capability. However, it has a different simulation and thermal calculation approach, which impedes the comparison of its intermodal validity (33).

The third classification, BIM-inherited EED, includes Autodesk GBS® and Graphisoft EcoDesigner® as its most popular applications. Concerning the inbuilt simulation engine of Autodesk GBS® in Autodesk Revit® and the fewer efforts required for the model preparation and simulation in BIM-inherited EED, the utilisation of this package has recently received much more attention. Nonetheless, Graphisoft EcoDesigner® has been significantly disregarded in the literature. Batueva and Mahdavi (2015) revealed that in the EcoDesigner environment, the different levels of model resolution are not supported for different design phases. In addition, more intelligence is required to provide a comparison capability for multiple design variants. Bentley AECOsim Energy Simulator® is another BIM authoring tool of this category (34).

In light of the BIM aware systems, a new group of simulation procedures, namely 'open BIM', can be established within the BIM-compatible and BIM-inherited EEDs to describe unique types of simulations. Open BIM simulations can be defined as building performance software-free prototypes. Open BIMs are applied with either data exchange protocols such as IFC and gbXML or BIM authoring tools databases to capture the required information and estimate the energy consumption of buildings (35). The first approach falls to the BIM-compatible, and the second way of working with BIM databases is grouped within BIM-inherited EED in terms of data exchange protocols. The openBIM simulation benefits from the higher level of flexibility in the prototype development but, in turn, suffers from the lack of a graphical interface. More importantly, its validity and accuracy cannot be rigorously evaluated. The open prototype development and applying personal judgements during data acquisition and calculation lead to the weak analogy fallacy and generate meaningless discourse in the analytical verification and intermodal comparisons.

3.2.6   *Interoperability*

According to the building energy performance simulation ethics by Williamson (26), interoperability is a clue in checking whether the states or outcomes of a simulation can be transferred to a context beyond its initial instinct or stage. Therefore, it can be expressed that the interoperability concept is more in the spotlight for BIM-compatible EED rather than the other two groups, considering the BIM-EED levels. As mentioned in the previous sections, in BIM-compatible EED, the trend is on BIM as the pivotal information repository for building energy analysis and the automatic development of the design model for energy simulators. The typical practice here is to transform the BIM model into input files for resolving the interoperability phenomenon and utilise IFC, gbXML and other sub-exchanges to forge an automated link between BIM authoring tools and energy performance software (36).

However, the inconsistency in the outcomes and functions of these tools is one of the problems damaging their reliability. Raslan and Davies (37) revealed the predictive deviation among IFC and gbXML through different pass/fail results concerning compliance with energy efficiency standards and guidelines. This procedure indicates the present routine of post-design performance evaluation. However, it does not develop the simulation combined with the design decision-making instead of the BIM-inherited EED (38). The technical problem in this procedure is the back-and-forth import and export of the model between the design and energy simulation software.

The interoperability methods used in BIM-EED mostly include IFC and gbXML as the first options and formats such as XML, IFC XML and DXF. Moon et al. (2011) stated that the wide range of BIM and energy simulation applications support the gbXML format in terms of BIM uptake. Still, Eastman et al. (2008) pronounced that IFC is regarded as the only public, non-proprietary and well-developed data protocol for the built environment thus far. IFC is employed more broadly than gbXML in advanced engineering informatics, as Cemesova, Hopfe and Rezgui (2013) stressed. However, it is claimed that gbXML is more understandable and applicable for building performance simulation software because it is grounded on the widely used XML standard (39).

Moreover, IFC includes more detailed and exhaustive semantics and syntax throughout the building disciplines and lifecycle from the database integrity perspective. Hence, it provides more reliable data export and import functionality. However, in a holistic view, gbXML is disadvantageous as it consists of a portion of building information. On the other hand, gbXML schema facilitate the better exchange of building geometry, material properties, HVAC information and building location, the principal information required for building energy simulation (39).

All in all, additional information, including the thermodynamic properties of building materials, systems for building services and occupants' programs, are not or only partially exchanged via these formats. For the most part, the geometry data are stored or imported, and the data modelling capabilities of gbXML and IFC formats are not fully implemented. Lately, among the BIM authoring tools, Revit is found to aid better the conversion of 3D meshes modelled in the design software into data applicable for energy analysis in simulative tools (40). However, manual adjustments should be implemented in the model and mesh during the back-and-forth procedure, especially where the mesh is complex, and the model is full-scale.

Existing energy simulation software (first and second generations) cannot provide conclusive parametric capabilities within the model and related objects in the discussed tools so far (9). For example, if an object such as a wall is edited or changed in an energy modelling tool, the other things, like roofs, doors or windows, are not revised automatically (41). In other words, parametric powers exerted in BIM are not implanted in the energy simulator (42). The same scenario turns out to be true on the flipside. Therefore, any alteration or correction in BIM design cannot be implemented simultaneously in the energy modelling software. Therefore, any modification from the energy simulator to BIM or vice versa must be manually conducted. This process is tedious, error-prone and time-consuming, especially for the models with a higher level of information (43).

3.2.7   Level of Details

The level of data and the richness of BIM models in all BIM-EED levels are defined via the concept of LoD, termed by the American Institute of Architects (AIA) in *AIA G202–2012 Building Information Modelling Protocol Form* (44). 'The level of detail describes the minimum dimensional, spatial, quantitative, qualitative and other data included in a model element to support the authorised uses associated with such LoD' (44). LoD was categorised in five levels (45):

- *LoD 100* (conceptual). The model element may be graphically represented in the model with a symbol or other generic representation but does not satisfy LoD 200. Information related to the model element (i.e. cost per square foot, the tonnage of HVAC, etc.) can be derived from other model elements.
- *LoD 200* (approximate geometry). The model element is graphically represented as a generic system, object or assembly with approximate quantities, size, shape, location and orientation. Non-graphic information may also be attached to the model element.
- *LoD 300* (precise geometry). The model element is graphically represented as a specific system, object or assembly in terms of quantity, size, shape, location, and orientation. Non-graphic information is also attached to the model element.
- *LoD 400* (fabrication). The model element is graphically represented as a specific system, object or assembly in terms of size, shape, location, quantity and orientation with detailing, fabrication, assembly and installation information. Non-graphic information is also attached to the model element.
- *LoD 500* (as-built). The model element is a field verified representation of size, shape, location, quantity, orientation and other semantic information.

Translating the above concepts into EED discourse, it can be stated that LoD 100, by including overall building massing, area, height and volume, can only be used to analyse the conceptual energy and building orientation. LoD 200 can be conducted in the next level if the BIM model gets equipped with some preliminary topological information, raw power and thermal studies and performance analysis of daylighting (39). LoD 300 contains the volume of information input in the models equivalent to the construction documentation. It can then be used for detailed energy consumption, lighting and solar radiation analyses and optimisation of systems. This LoD generates the most profound effect on the building performance when it is particularly performed early in the design stage (conceptual, schematic and design development phases) (46). The higher levels of LoD 400 and LoD 500 correspond to the construction and operation phases of the building projects' lifecycle.

The current application of LoDs in the BIM-EED presents the absolute majority of LoD 100 and 200 for the BIM models. However, LoD 300 is completely neglected in the domain. Of course, there is a recent trend in BIM-EED in which research studies have been more inclined to enhance the LoD of BIM models and shift from LoD 100/200 into LoD 300. One of the reasons for the low LoD (100 and 200) in BIM-EED is that significant time is required for learning and achieving proficiency in using BIM authoring tools to enhance their LoD. Practitioners regularly ask for assistance and consultation from the experts in their network and rely on 'pass[ing] down knowledge from person to person' (47).

In addition, reviewing the procedural instruction of these applications indicates a lack of comprehensive guidelines in advising the academic body to create and implement BIM models in the appropriate LoD applicable for an accurate energy assessment (48). For instance, in the EcoDesigner environment, the level of model details cannot be exactly matched to the level and resolution of the query. Thus, the requirements for model input remain the same, independent of what level of the analysis is intended. Therefore, the tool does not support varying levels of model resolution for different design phases (49). Figure 3.2 represents different LoDs and their embedded information concerning the BIM-EED categories.

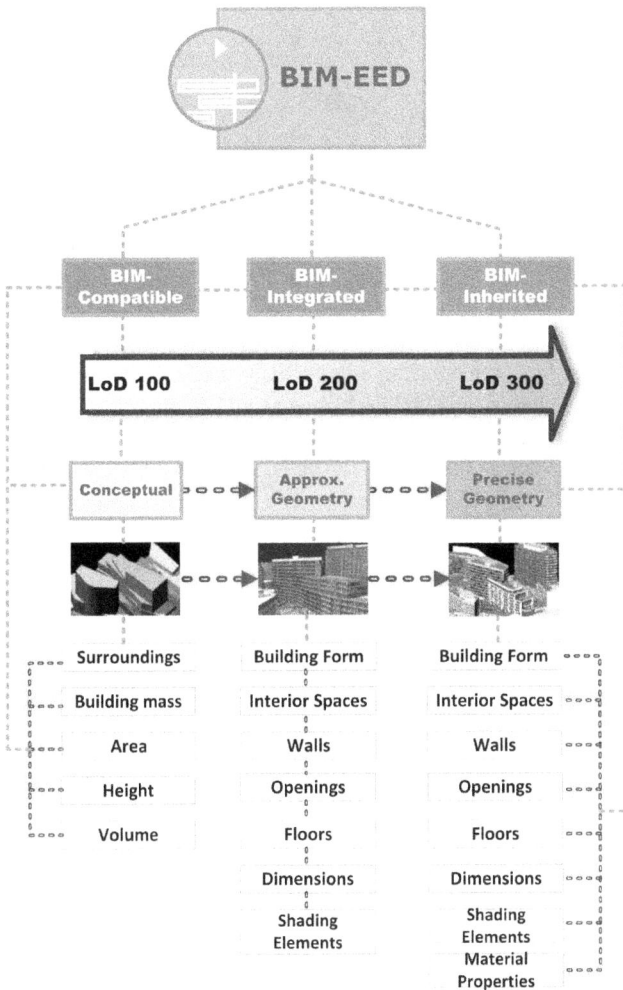

Figure 3.2 LoD and the Contained Information in BIM-EED
Source: Adapted from (44, 46)

3.3   Themes and Gaps

3.3.1   Themes

The academic studies on BIM-EED, for the most part, have targeted case studies as their main research themes. Such studies mostly indicate the various technical aspects of BIM authoring tools and their interactions with energy analysis packages. There are also theoretical and conceptual models that generally introduce a foundation or groundwork for BIM-compatible, e.g. (50–52), BIM-integrated, e.g. (53–55) and BIM-inherited EED, e.g. (23, 56, 57). However, survey-based research studies are the theme least taken into account by the scientific community. This fact brings to light the lack of large-scale surveys to understand the perception of experts and practitioners toward the BIM-EED domain. Besides, there should be a shift from case studies to industry-scale investigations to address the practice shortcomings and alleviate the functional and industrial performance doubts about BIM-EED (Figure 3.3).

3.3.2   Outcomes

The academic BIM-EED outcomes are mostly concentrated on study analysis-based outcomes. This category of outcomes mainly arises from the case study research themes. It provides an analytical account of employing BIM-EED levels and the associated barriers and procedures, e.g. (49, 58, 59). The next type, tool/ system prototype, generally results from both conceptual and case study themes and can be regarded as the cutting edge of BIM-EED literature.

For example, Park et al. (2012) proposed a BIM-based energy performance assessment system for an energy performance index in Korea applying open BIM, visual basic and Microsoft Access® to the BIM-compatible class. Chen and Gao (60) developed a multi-objective genetic algorithm approach to optimise building energy performance grounded on BIM-integrated EED. In like manner, Hu et al. (61) presented a system porotype of energy management of large airport projects using a multi-scale BIM model within the BIM-inherited EED level. The next outcomes, frameworks including theoretical and conceptual models, are generated

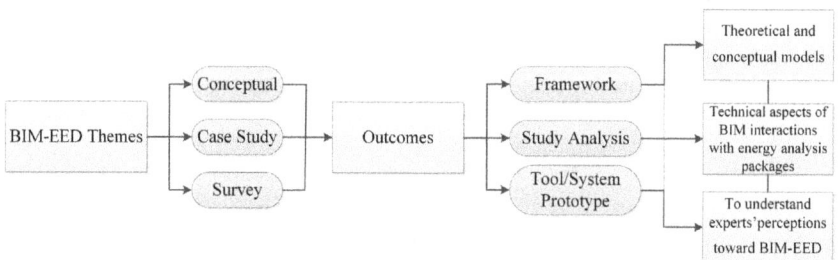

Figure 3.3 Themes and Outcomes of BIM-EED

from conceptually based themes and present logic or rules of establishing the concepts for BIM-compatible (62, 63), BIM-integrated (64, 65) and BIM-inherited EED (34, 66) (Figure 3.3).

3.3.3 Gaps

This section reveals the scant regard of the academic body for knowledge of BIM-EED because of gap-spotting modes, namely, confusion, neglect and application. The gap analysis is the most prevalent way of developing research questions in empirical studies, but it is often applied intuitively in built environment research. This can be due to a lack of awareness about the structured manner of gap problematisation (67). There are three particular modes of gap analysis, confusion, neglect and application spotting, and these are elaborated based on a seminal study by Sandberg and Alvesson (68) to present the theoretical framework of the gap analysis for this section:

- *Confusion.* The identification of any sort of confusion in the existing body of knowledge is the main focus here. Previous studies exist on the subject; however, available explanation, interpretation or evidence is antagonistic. Therefore, the research question is pointed to for figuring out the specified confusion in the literature and its explanation. The practical approach of such an analysis looks for competing explanations in the body of knowledge.
- *Neglect.* The most dominant method of developing research questions is to spot something neglected in the body of knowledge. The aim is to find a realm of research where virgin territory makes imperative for scholars to expand their knowledge and call for more investigations. This mode of gap analysis is further categorised into three versions: spotting an overlooked area, an under-researched area or a lack of empirical support.
- *Application.* The third mode of gap analysis spots an area in the existing body of knowledge. Again, a shortage of an exact theory or a distinct perspective in a particular research domain is targeted. The task here is to identify the studies solely focused on some case studies and their applications. These are attributed to the limited contribution to the body of knowledge or the lack of presenting an alternative viewpoint to expedite the scientific community's understanding of that specific research question. Typically, advocates of the application spotting claim that specific frameworks and/or research pathways should be developed so that the body of literature is extended or complemented in some way (68).

3.3.3.1 Confusion

There is confusion between BIM as a tool and BIM as a process. BIM can be interpreted from different angles, such as a tool providing nD modelling, an activity developing a building information model or a lifecycle management and coordination building project (69). According to Succar (70), BIM integrates processes, technologies and policies that interact with each other and generate a methodology to manage the information and design systems in digital schema over the building lifecycle. However, as identified by (71), the BIM model is a digital representation of functional and physical characteristics. Since there are competing explanations on BIM interpretation, this vice indulges in BIM-EED adoption levels confusion.

As a consequence, studies of Jašek, Česelský (72), Kim, Jin and Choi (73) and Reeves, Olbina and Issa (32) were trapped in losing focus on BIM-compatible, BIM-integrated or BIM-inherited adoption levels. Likewise, Choi, Shin (74) and Ham and Golparvar (75) were confused with targeting the design or operational phase of the project lifecycle and the outputs applicable for BIM authoring tools. Therefore, BIM-EED confusions can be due to lack of appreciation of the dimensions associated with the BIM interpretation maturity. Notably, the conceptual definition is the basis of further research and discussions on the subject. Therefore, any ambiguity in this area could directly result in vagueness of all issues regarding BIM-EED.

3.3.3.2   Neglect

The neglected realms of BIM-EED can be broken down into the overlooked, under-researched and lack of empirical supports. Despite being extensively investigated in the literature, interoperability is still a fundamental issue in the BIM-EED domain and, more specifically, for BIM-compatible EED. When analysing interoperability, data exchange specifications such as temporal data management, transaction management and dynamic synchronisation capabilities should be studied. However, to some extent, the literature is neglected and overlooked toward the interoperability phenomenon (76, 77). This gap is widened by using imprecise and unsuitable LoD in the body of knowledge. For example, Charalambides (78) and Jansson, Schade and Olofsson (79) utilised BIM models containing LoD 100 and even LoD 200. Thus, these studies do not seem reliable enough for pinpointing the interoperability and technical information aspects of BIM-EED due to the poor level of complexity and data organisation. Eventually, the over-simplified energy calculation denotes another overlooked area of the topic, exemplified by two pieces of publication (80, 81).

As the second subgroup of the neglect gaps, the under-researched areas are primarily around data analytics function and lack of a graphical user interface. In line with the increasing use of linked data analytics, as expressed by Bradley, Li (82), AI can assist EED through facilitating tests of the design parameters for 'what if' scenarios. So, it can be expected that such platforms could be successfully applied as a decision-making tool. Nevertheless, by reviewing studies such as Chen and Gao (60) and Gokce (83), it is inferred that the potential of inbuilt data analytics, integrated machine learning and optimisation procedures have been ignored in the field. Exploiting an appropriate graphical user interface level also leads to enhanced utilisation of BIM adoption into the EED. It facilitates the usage of ontologies and structured data formats within BIM.

Neglecting the integrated graphical interface mostly belongs to the openBIM studies (23, 84–86).

The lack of empirical support presents the third subgroup of the neglect gaps and constitutes a considerable part. Providing proof of concept guarantees the feasibility of an idea or verifies a concept or theory that has practical potential. However, the analysis shows that most of the theoretical or conceptual frameworks that were supposed to present the nature of the BIM-EED application and integration have not had empirical support. This fact raises scepticism about the extent to which conceptual-based research themes by Grinberg and Rendek (87) and Gudnason, Katranuschkov (88) could be employed in the industry-scaled BIM-EED problems.

3.3.3.3   *Application*

This mode of the gap in the BIM-EED is solely derived from the case study relevant research themes and, for the most part, is rooted in study analysis-based outcomes. The technical characteristics of this gap include concentrating on the simulation of building energy performance (89–91) and treating BIM as a tool rather than a process (92, 93). More remarkably, this type of research delivers limited contributions to the theoretical and conceptual foundations of the body of BIM-EED knowledge. Accordingly, it affects the academic body's fundamental research and development efforts. This gap may lie in the dominance of the technology-oriented view among contributors and the engineering-based scholarships in BIM-EED. On the other hand, it may be due to bias toward the quantitative aspects of BIM-EED and the absence of sufficient awareness about the depth of the knowledge and systematic approaches.

3.4   The Future of BIM-EED

As a result of the comprehensive analysis reflected thus far, and taking into account the viewpoints mentioned, the following key points outline and determine the grounds for the future of the BIM-EED domain as presented in Figure 3.4 and summarised below:

1   The interpretative confusion of BIM as a practice and tool or BIM as the knowledge and process should be investigated, and the link with BIM-EED needs to be uncovered. Priority could be given more to BIM as the process because of the dimensional appreciation of BIM implementation levels in EED.

2   A shift from low to high levels of BIM adoption is imperative. Such adoption could be in line with the classification of BIM-compatible, BIM-integrated and BIM-inherited EED introduced in this book.

3   There are numerous metaphorical studies for BIM-compatible and BIM-integrated EED applications, yet, BIM-inherited EED tools such as Autodesk GBS®, Graphisoft EcoDesigner® and Bentley AECOsim® require comparative analysis among them to demonstrate the functionality and performance of their energy simulators.

4   The available energy simulation packages suffer from a broad spectrum of drawbacks, ranging from lack of graphical user interface and unreliable data exchanges to the poor thermal engine. Therefore, researchers should continue the research and development pathway to find more innovative ways to resolve the identified issues.

5   The interoperability concept's functions, comprehensiveness, and data inclusion levels in particular could be a set of paramount areas for BIM-compatible EED. For example, poor data processing load denotes the lack of attention to the different rates of uncertainties, so the design simulation cannot be fully realistic.

6   Even if applying the improved data formats of IFC and gbXML, the interoperability phenomenon dictates restrictive circumstances on the EED development. Normally, these schemas are less robust than those of BIM aware applications. Hence, it is recommended to focus on the inbuilt interoperability within BIM-inherited EED because it automatically resolves interoperability problems.

7   LoD improvement from LoD 100 and LoD 200 to LoD 300 should be undertaken as a routine task in BIM-EED studies. It was observed that the overlooked LoDs lead to superficial assessment of building energy performance.

8   The conversion and complex mesh development capabilities of full-scale BIM models should be analysed in conjunction with the semantic and syntax of IFC and gbXML. However, this research agenda makes sense when the higher levels of LoD are implemented.

9   An especially designated standard, guideline or specification should be developed to detail the LoD and required information for an energy efficient built environment. In addition, this protocol should be accessible for energy and building experts, practitioners and academics to advise them of the appropriate class of LoD for BIM-EED adoption.

10 The intelligence agent of BIM-EED should be built up. Furthermore, substantial improvements such as active design decision-making, advisory platforms and energy optimisation recommenders should be developed and embedded into BIM for assisting designers in choosing the most optimised options.

11 Data analytics and big data handling in the energy efficiency problems should be exploited for leveraging the parametric definition inherited in BIM. Henceforth, it seems that BIM-inherited EED is the best choice for testing the inbuilt AI applications.

12 Large and industrial-scale surveys could benefit the scientific community in terms of understanding the current gaps and future needs of BIM-EED.

13 The sheer volume of BIM-EED case study-based research should be established on the theoretical frameworks.

14 Validating conceptual frameworks through applying proof of concept will reasonably raise the reliability of the theoretical knowledge of BIM-EED.

15 Last but not least, industrially verified tools and system prototypes open a new window of opportunity for BIM-EED to be widely applied and become more cost-efficient for end-users.

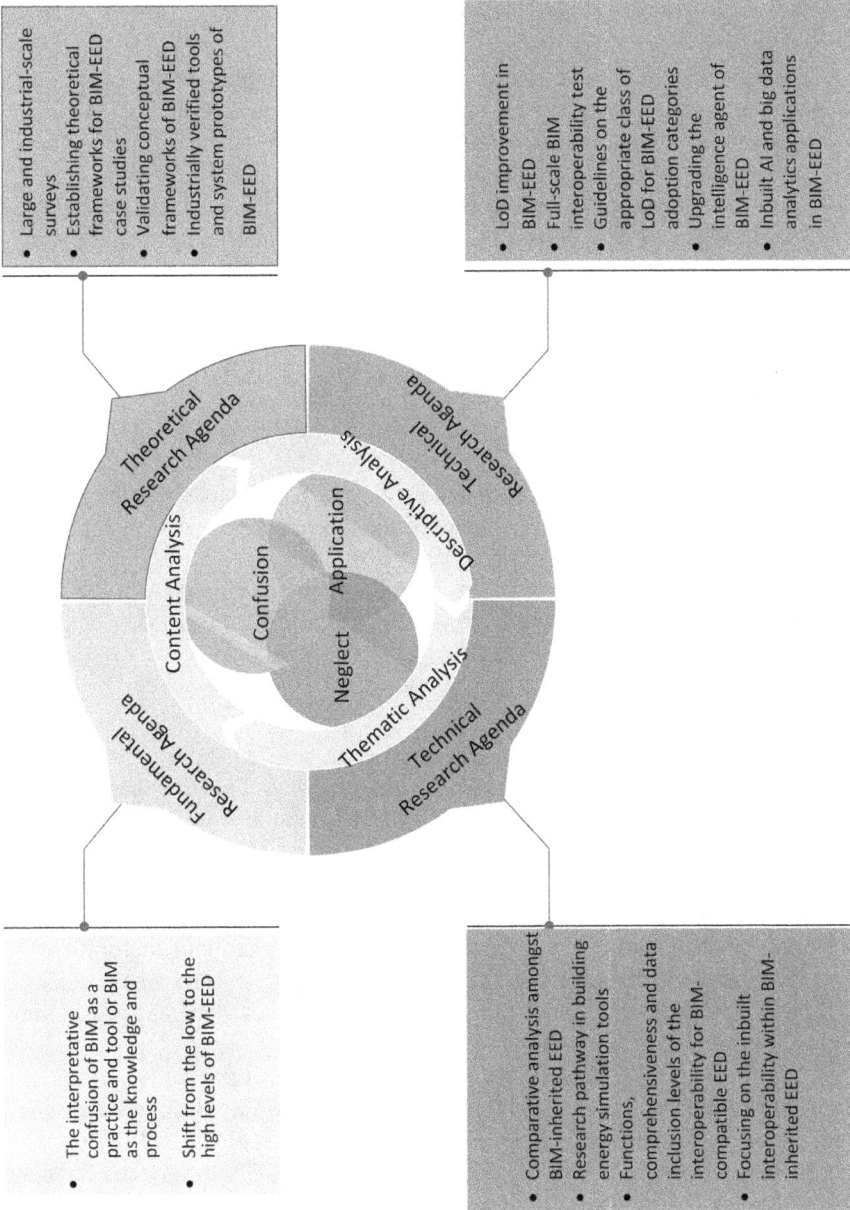

Theoretical Research Agenda

- Large and industrial-scale surveys
- Establishing theoretical frameworks for BIM-EED case studies
- Validating conceptual frameworks of BIM-EED
- Industrially verified tools and system prototypes of BIM-EED

Technical Research Agenda

- LoD improvement in BIM-EED
- Full-scale BIM interoperability test
- Guidelines on the appropriate class of LoD for BIM-EED adoption categories
- Upgrading the intelligence agent of BIM-EED
- Inbuilt AI and big data analytics applications in BIM-EED

Fundamental Research Agenda

- The interpretative confusion of BIM as a practice and tool or BIM as the knowledge and process
- Shift from the low to the high levels of BIM-EED

Technical Research Agenda

- Comparative analysis amongst BIM-inherited EED
- Research pathway in building energy simulation tools
- Functions, comprehensiveness and data inclusion levels of the interoperability for BIM-compatible EED
- Focusing on the inbuilt interoperability within BIM-inherited EED

Content Analysis — Confusion — Application — Neglect — Descriptive Analysis — Thematic Analysis

Figure 3.4 The Future of BIM-EED

3.5   Implications

The developed analysis has clear implications for this book and the BIM-EED concept, which can be summarised as below:

- BIM-inherited EED is of particular focus for the developed project in this book. As realised, BIM-inherited EED is on the highest level of BIM adoption into EED. Hence, to achieve the most out of BIM potential, the AI-based active BIM will be developed on a BIM-inherited EED basis.
- The thematic and research outcomes analysis indicates that a consolidated approach should be designed for BIM-EED research. This fact is in light of developing a conceptual framework and verifying its functionality through a case study. Therefore, the project should comprise an established and robust integration of the theoretical background, framework development and validation via the case study.
- Out of the various BIM-inherited EED applications, Autodesk GBS-Revit® is the chosen package to analyse and test the validity and reliability of the framework in this book.
- To enhance the precision, depth and inclusion of BIM models in the framework, LoD 300 would be the target of the development and verification as the topmost LoD for the design stage.
- The parametric definition of BIM will be substantiated through AI algorithms to enable data analytics and energy optimisation inbuilt in BIM.

3.6   Summary

This chapter contributed to the body of knowledge on BIM and energy efficient built environment, particularly BIM-EED, in several ways. First, the chapter drew a picture of the current state of the art of BIM-EED adoption. It was then grounded on the systematic, thematic and gap analysis frameworks to provide a view on the theme from a meta-perspective and give an identity to the available academic literature of the domain. This included predominant research trends, preferences and potential content areas, including BIM-EED adoption levels, simulation software, interoperability and the LoD of BIM models. Finally, the chapter also provided evidence for anecdotal statements in previous studies about the gaps within the body of knowledge on the topic.

On top of that, the established benchmarking framework provides a sound basis for future investigators. This helps identify the themes, outcomes and spot areas that need further research. Finally, the future agenda of BIM-EED directed this research to strategically place the AI-enabled BIM-EED study and form a solid basis for sharing knowledge and collaboration in the area. The clear message is that the time has come to move beyond the current debates in the passive BIM-EED and shift to the active BIM-inherited EED to rectify the outlined drawbacks.

References

1 Pacheco R, Ordóñez J, Martínez G. Energy efficient design of building: A review. Renewable and Sustainable Energy Reviews. 2012;16(6):3559–73.

2 Ekici BB, Aksoy UT. Prediction of building energy needs in early stage of design by using ANFIS. Expert Systems with Applications. 2011;38(5):5352–8.

3 Soares N, Bastos J, Pereira LD, Soares A, Amaral A, Asadi E, et al. A review on current advances in the energy and environmental performance of buildings towards a more sustainable built environment. Renewable and Sustainable Energy Reviews. 2017;77:845–60.

4 Shi X, Tian Z, Chen W, Si B, Jin X. A review on building energy efficient design optimization from the perspective of architects. Renewable and Sustainable Energy Reviews. 2016;65:872–84.

5 Shi X, Yang W. Performance-driven architectural design and optimization technique from a perspective of architects. Automation in Construction. 2013;32(Supplement C):125–35.

6 Larsson J, Jansson J, Olofsson T, Simonsson P, editors. Increased innovation through change in early design procedures. 19th IABSE Congress, Challenges in Design and Construction of an Innovative and Sustainable Built Environment; 2016.

7 Machairas V, Tsangrassoulis A, Axarli K. Algorithms for optimization of building design: A review. Renewable and Sustainable Energy Reviews. 2014;31:101–12.

8 Asadi E, da Silva MG, Antunes CH, Dias L, Glicksman L. Multi-objective optimization for building retrofit: A model using genetic algorithm and artificial neural network and an application. Energy and Buildings. 2014;81:444–56.

9 Ahn K-U, Kim Y-J, Park C-S, Kim I, Lee K. BIM interface for full vs. semi-automated building energy simulation. Energy and Buildings. 2014;68:671–8.

10 Aquino EV. Predicting building energy performance: Leveraging BIM content for energy efficient buildings; 2013.

11 Yalcinkaya M, Singh V. Patterns and trends in Building Information Modeling (BIM) research: A latent semantic analysis. Automation in Construction. 2015;59:68–80.

12 Lu Y, Li Y, Skibniewski M, Wu Z, Wang R, Le Y. Information and communication technology applications in architecture, engineering, and construction organizations: A 15-year review. Journal of Management in Engineering. 2014;31(1):401–10.

13 Eleftheriadis S, Mumovic D, Greening P. Life cycle energy efficiency in building structures: A review of current developments and future outlooks based on BIM capabilities. Renewable and Sustainable Energy Reviews. 2017;67:811–25.

14 Foucquier A, Robert S, Suard F, Stéphan L, Jay A. State of the art in building modelling and energy performances prediction: A review. Renewable and Sustainable Energy Reviews. 2013;23:272–88.

15 Negendahl K. Building performance simulation in the early design stage: An introduction to integrated dynamic models. Automation in Construction. 2015;54:39–53.

16 Hammersley M. On 'systematic' reviews of research literatures: A 'narrative' response to Evans & Benefield. British Educational Research Journal. 2001;27(5):543–54.

17 Wang Z, Srinivasan RS. A review of artificial intelligence based building energy use prediction: Contrasting the capabilities of single and ensemble prediction models. Renewable and Sustainable Energy Reviews. 2017;75:796–808.

18 Day GS, Schoemaker PJ, Gunther RE. Wharton on managing emerging technologies: John Wiley & Sons; 2004.

19 Crawley DB, Hand JW, Kummert M, Griffith BT. Contrasting the capabilities of building energy performance simulation programs. Building and Environment. 2008;43(4):661–73.

20 Wang H, Zhai Z. Advances in building simulation and computational techniques: A review between 1987 and 2014. Energy and Buildings. 2016;128:319–35.

21 Watson A. Digital buildings: Challenges and opportunities. Advanced Engineering Informatics. 2011;25(4):573–81.

22 Kim H, Shen ZH, Kim I, Kim K, Stumpf A, Yu J. BIM IFC information mapping to Building Energy Analysis (BEA) model with manually extended material information. Automation in Construction. 2016;68:183–93.

23 Banihashemi S, Ding G, Wang J, editors. Developing a framework of artificial intelligence application for delivering energy efficient buildings through active BIM. COBRA 2015, RICS; 2015.

24 Mencarini R. Big data and BIM: Turning model information into insight. Fall BIMForum; Dallas, TX2014.

25 Attia SGM. State of the art of existing early design simulation tools for net zero energy buildings: A comparison of ten tools. UCL; 2011.

26 Williamson TJ. Predicting building performance: The ethics of computer simulation. Building Research & Information. 2010;38(4):401–10.

27 Augenbroe G, editor. Building simulation trends going into the new millennium. In Building simulation. IBPSA; 2001.

28 Crawley DB, Lawrie LK, Winkelmann FC, Buhl WF, Huang YJ, Pedersen CO, et al. EnergyPlus: Creating a new-generation building energy simulation program. Energy and Buildings. 2001;33(4):319–31.

29 Ryan EM, Sanquist TF. Validation of building energy modeling tools under idealized and realistic conditions. Energy and Buildings. 2012;47:375–82.

30 Attia SGM, Beltrán L, Hensen JL, De Herde A. 'Architect friendly': A comparison of ten different building performance simulation tools. International Building Performance Simulation Association. 2009;1.

31 Azhar S, Brown J, Farooqui R, editors. BIM-based sustainability analysis: An evaluation of building performance analysis software. Proceedings of the 45th ASC Annual Conference; 2009.

32 Reeves T, Olbina S, Issa RR. Guidelines for using building information modeling for energy analysis of buildings. Buildings. 2015;5(4):1361–88.

33 Somboonwit N, Sahachaisaeree N. Healthcare building: Modelling the impacts of local factors for building energy performance improvement in Thailand. In: Abbas MY, Bajunid AFI, Azhari NFN, editors. Ace-Bs 2012 Bangkok. Procedia Social and Behavioral Sciences. 50, Elsevier Science Bv, Amsterdam; 2012. p. 549–62.

34 Kota S, Kim JB, Yan W, Stipo FJF, Alcocer JLB, Haberl JS, et al., editors. Development of a reference building information model for thermal model compliance testing – Part I: Guidelines for generating thermal model input files. ASHRAE Transactions; 2016.

35 Choi J, Kim M, Kim I. A methodology of mapping interface for energy performance assessment based on open BIM. Proceedings of the 20th International Conference on Computer-Aided Architectural Design Research in Asia (CAADRIA 2015): Emerging Experiences in the Past, Present and Future of Digital Architecture. 2015:417–26.

36 Bahar YN, Pere C, Landrieu J, Nicolle C. A thermal simulation tool for building and its interoperability through the Building Information Modeling (BIM) platform. Buildings. 2013;3(2):380–98.

37 Raslan R, Davies M. Results variability in accredited building energy performance compliance demonstration software in the UK: An inter-model comparative study. Journal of Building Performance Simulation. 2010;3(1):63–85.

38 El Hjouji ME, Khaldoun A, editors. Thermal auditing of buildings: Essential step towards designing energy efficient houses. Renewable and Sustainable Energy Conference (IRSEC), 2013 International; March 7–9, 2013.

39 Nasyrov V, Stratbücker S, Ritter F, Borrmann A, Hua S, Lindauer M. Building information models as input for building energy performance simulation – the current state of industrial implementations. eWork and eBusiness in Architecture, Engineering and Construction, ECPPM 2014. 2014. p. 479.

40 Hitchcock RJ, Wong J, editors. Transforming IFC architectural view BIMs for energy simulation: 2011. Proceedings of Building Simulation; 2011.

41 Hensen JL, Lamberts R. Building performance simulation for design and operation: Routledge; 2012.

42 Cho YK, Alaskar S, Bode TA, editors. BIM-integrated sustainable material and renewable energy simulation. Proceedings of the Construction Research Congress; 2010.

43 Asl MR, Bergin M, Menter A, Yan W. BIM-based parametric building energy performance multi-objective optimization. Fusion, Proceedings of the 32nd International Conference on Education and Research in Computer Aided Architectural Design in Europe, Northumbria University, Newcastle upon Tyne, UK; September 2014. p. 455–64.

44 AIA. AIA document G202™ – 2012. Project Building Information Modeling Protocol Form; 2012.

45 NATSPEC. BIM and LoD: Building information modeling and level of development. Construction Information System; 2013.

46 Aksamija A. BIM-based building performance analysis: Evaluation and simulation of design decisions. Proceedings of the 2012 ACEEE Summer Study on Energy Efficiency in Buildings; 2012.

47 Tupper K, Franconi E, Chan C, Hodgin S, Buys A, Jenkins M, editors. Building energy modeling: Industry-wide issues and potential solutions. Proceedings of the 12th Conference of International Building Performance Simulation Association; 2011.

48 Kavgic M, Mavrogianni A, Mumovic D, Summerfield A, Stevanovic Z, Djurovic-Petrovic M. A review of bottom-up building stock models for energy consumption in the residential sector. Building and Environment. 2010;45(7):1683–97.

49 Batueva E, Mahdavi A. Assessment of a computational design environment with embedded simulation capability. Mahdavi A, Martens B, Scherer R, editors. CRC Press-Taylor & Francis Group; 2015. p. 197–202.

50 Thomas D, Schlueter A. Customizable continuous building simulation using the design performance toolkit and Kepler scientific workflows. Ework and Ebusiness in Architecture, Engineering and Construction. 2012:17–21.

51 Kim S, Woo JH. Analysis of the differences in energy simulation results between building information modeling (BIM)-based simulation method and the detailed simulation method. In: Jain S, Creasey R, Himmelspach J, editors. Proceedings of the 2011 Winter Simulation Conference, Winter Simulation Conference Proceedings, IEEE, New York; 2011. p. 3545–56.

52 Kim H, Anderson K, editors. Energy simulation system using BIM (Building Information Modeling). Congress on Computing in Civil Engineering, Proceedings; 2011.

53 Majid MZBA, Marsono AKB, Golzarpoor H, Sadi MK, editors. A comparative study of energy consumption and carbon emission analyses between RNC And NRNC based

on BIM technology. CRIOCM 2010 – International Symposium on Advancement of Construction Management and Real Estate 'Towards Sustainable Development of International Metropolis'; 2010.

54   Hu M, editor. Performance-based design. Open Systems: Proceedings of the 18th International Conference on Computer-Aided Architectural Design Research in Asia, CAADRIA 2013; 2013.

55   Cao J, Wimmer R, Thorade M, Maile T, O'Donnel J, Rädler J, et al., editors. A flexible model transformation to link BIM with different Modelica libraries for building energy performance simulation. Proceedings of the 14th IBPSA Conference; 2015.

56   Inyim P, Rivera J, Zhu YM. Integration of building information modeling and economic and environmental impact analysis to support sustainable building design. Journal of Management in Engineering. 2015;31(1).

57   Hu S, Corry E, Curry E, Turner WJN, O'Donnell J. Building performance optimisation: A hybrid architecture for the integration of contextual information and time-series data. Automation in Construction. 2016;70:51–61.

58   Stumpf A, Kim H, Jenicek E, editors. Early design energy analysis using BIMs (building information models). Building a Sustainable Future: Proceedings of the 2009 Construction Research Congress; 2009.

59   Tahmasebi MM, Banihashemi S, Hassanabadi MS, editors. Assessment of the variation impacts of window on energy consumption and carbon footprint. Procedia Engineering; 2011.

60   Chen D, Gao Z, editors. A multi-objective generic algorithm approach for optimization of building energy performance. Congress on Computing in Civil Engineering, Proceedings; 2011.

61   Hu ZZ, Zhang JP, Yu FQ, Tian PL, Xiang XS. Construction and facility management of large MEP projects using a multi-scale building information model. Advances in Engineering Software. 2016;100:215–30.

62   Motawa I, Carter K. Sustainable BIM-based evaluation of buildings. In: Pantouvakis JP, editor. Selected Papers from the 26th Ipma. Procedia Social and Behavioral Sciences. 74. Amsterdam: Elsevier Science Bv; 2013. p. 419–28.

63   Yu N, Jiang Y, Luo L, Lee S, Jallow A, Wu D, et al., editors. Integrating BIM server and OpenStudio for energy efficient building. Proceedings of 2013 ASCE International Workshop on Computing in Civil Engineering; 2013.

64   Remmen P, Cao J, Ebertshäuser S, Frisch J, Lauster M, Maile T, et al., editors. An open framework for integrated BIM-based building performance simulation using Modelica. 14th International Conference of IBPSA: Building Simulation 2015, BS 2015, Conference Proceedings; 2015.

65   Zhu S, Tu Y. Design study on green sports building based on BIM. Open Cybernetics and Systemics Journal. 2015;9:2479–83.

66   Lee YG. Developing a design supporting system in the real-time manner for low-energy building design based on BIM. In: Stephanidis C, editor. HCI International 2016: Posters' Extended Abstracts: 18th International Conference, HCI International 2016 Toronto, Canada, July 17–22, 2016 Proceedings, Part II. Springer International Publishing, Cham; 2016. p. 503–6.

67   Fellows RF, Liu AM. Research methods for construction: John Wiley & Sons; 2015.

68   Sandberg J, Alvesson M. Ways of constructing research questions: Gap-spotting or problematization? Organization. 2010;18(1):23–44.

69   Alsaadi A. sustainability and BIM the Role of Building Information Modeling to enhance energy efficiency analysis. University of Salford; 2014.

70 Succar B. Building information modelling framework: A research and delivery foundation for industry stakeholders. Automation in Construction. 2009;18(3):357–75.

71 Stumpf AL, Kim H, Jenicek EM. Early design energy analysis using building information modeling technology. DTIC Document; 2011.

72 Jašek M, Česelský J, Vlček P, Černíková M, Beránková E. Application of BIM process
by the evaluation of building energy sustainability. In: Advanced materials research.
Trans Tech Publications Ltd.; 2014. p. 7–10.

73 Kim I-H, Jin J, Choi J-S. A study on open BIM based building energy evaluation based
on quantitative factors. Korean Journal of Computational Design and Engineering.
2010;15(4):289–96.

74 Choi J, Shin J, Kim M, Kim I. Development of openBIM-based energy analysis software to improve the interoperability of energy performance assessment. Automation in
Construction. 2016;72:52–64.

75 Ham Y, Golparvar-Fard M. Mapping actual thermal properties to building elements in
gbXML-based BIM for reliable building energy performance modeling. Automation in
Construction. 2015;49:214–24.

76 Bank LC, McCarthy M, Thompson BP, Menassa CC. Integrating BIM with system
dynamics as a decision-making framework for sustainable building design and operation. Proceedings of the First International Conference on Sustainable Urbanization
(Icsu 2010). 2010. p. 15–23.

77 Delghust M, Strobbe T, De Meyer R, Janssens A, editors. Enrichment of single-zone
EPB-data into multi-zone models using BIM-based parametric typologies. 14th International Conference of IBPSA: Building Simulation 2015, BS 2015, Conference Proceedings; 2015.

78 Charalambides J, editor. Improving energy efficiency in building through automated
computer design process. Building a Sustainable Future: Proceedings of the 2009 Construction Research Congress; 2009.

79 Jansson G, Schade J, Olofsson T. Requirements management for the design of energy
efficient buildings. Journal of Information Technology in Construction.
2013;18:321–37.

80 Stojanovic V, Falconer R, Blackwood D, Paterson G, Fleming M, Bell S, editors. Interactive visualisation of heat loss and gain for early-stage energy appraisal of the built
environment. Proceedings 30th Annual Association of Researchers in Construction
Management Conference, ARCOM 2014; 2014.

81 Jones K, Gerdelan A, Cochero A, Lewis D, editors. Usability evaluation of a web-based
activity modelling tool for improving accuracy of predictive energy simulations. 14th
International Conference of IBPSA – Building Simulation 2015, BS 2015, Conference
Proceedings; 2015.

82 Bradley A, Li H, Lark R, Dunn S. BIM for infrastructure: An overall review and con-
structor perspective. Automation in Construction. 2016;71:139–52.

83 Gokce HU, Gokce KU. Integrated system platform for energy efficient building operations. Journal of Computing in Civil Engineering. 2014;28(6).

84 Verstraeten R, Pauwels P, De Meyer R, Meeus W, Van Campenhout J, Lateur G. IFC-
based calculation of the Flemish energy perfomance standard. In: Zarli A, Scherer R,
editors. eWork and eBusiness in Architecture, Engineering and Construction. CRC
Press-Taylor & Francis Group; 2009. p. 437–43.

85 Park J, Park J, Kim J, Kim J. Building information modelling based energy performance assessment system: An assessment of the energy performance index in Korea.
Construction Innovation. 2012;12(3):335–54.

86 Pour Rahimian F, Chavdarova V, Oliver S, Chamo F, Potseluyko Amobi L. OpenBIM-Tango integrated virtual showroom for offsite manufactured production of self-build housing. Automation in Construction. 2019;102:1–16.

87 Grinberg M, Rendek A, editors. Architecture & energy in practice: Implementing an information sharing workflow. Proceedings of BS 2013: 13th Conference of the International Building Performance Simulation Association; 2013.

88 Gudnason G, Katranuschkov P, Balaras C, Scherer RJ, editors. Framework for interoperability of information resources in the building energy simulation domain. Computing in Civil and Building Engineering: Proceedings of the 2014 International Conference on Computing in Civil and Building Engineering; 2014.

89 Douglass CD, Leake JM, editors. Instructional modules demonstrating building energy analysis using a building information model. ASEE Annual Conference and Exposition, Conference Proceedings; 2011.

90 Shakouri M, Banihashemi S. Developing an empirical predictive energy-rating model for windows by using Artificial Neural Network. International Journal of Green Energy. 2012. DOI: 10.1080/15435075.2012.738451.

91 Alam J, Ham JJ. Towards a BIM-based energy rating system. Proceedings of the 19th International Conference on Computer-Aided Architectural Design Research in Asia (CAADRIA 2014): Rethinking Comprehensive Design: Speculative Counterculture. 2014:285–94.

92 Ferrari PC, Silva NF, Lima EM. Building information modeling and interoperability with environmental simulation systems. In: Sobh T, editor. Innovations and Advances in Computer Sciences and Engineering. Springer-Verlag; 2010. p. 579–83.

93 Li YF, Wang HC, Zhao M, Pan WY. Analyses on the BIM technology using in the design of green village buildings. In: Applied mechanics and materials. Trans Tech Publications Ltd.; 2014. p. 1758–62.

4 Building Energy Parameters

4.1 Building Energy Parameters

At the building level, when designing with the purpose of energy performance optimisation, a large number of factors and variables can be considered. But some constraints, like time, the scope of the project and the availability of resources and facilities, may convince the researchers to narrow down the variables into a manageable number. The quest for optimum energy consumption requires a coherent implementation of factors that together optimise the performance of the whole building system (1). In the literature, a huge number of variables and parameters have been practised and different types of categorisations have been introduced.

From the design perspective, the parameters influencing the energy consumption of buildings can be categorised into non-design and design factors (2). In this categorisation, occupant behaviour, climate and indoor environmental quality have been classified as non-design factors and building layout, physical properties and HVAC related variables are labelled as design factors (2). In another categorisation by Park, Park (3), the factors have been classified based on the type of information they contain, in which geometric information relates to the 3D form of the building. Semantic information implies the properties of components like U-values of walls and topological information describe the dependencies of these components (3).

Considering the different types of categorisations (2–4), generally, the significant design and construction parameters influencing buildings' energy consumption can be classified into four categories (4):

1   Physical properties and building envelop
2   Building layout
3   Occupant behaviour
4   HVAC and appliances

4.1.1 Physical Properties and Building Envelop

Building envelop is the key area where thermal losses happen, and so, it is very significant in contribution to energy loss. It is a major path in thermal loss due to heat transfer in winter and an excessive solar heat gain area in summer (4).

DOI: 10.1201/9781003207658-4

For this reason, a building's heating and cooling demands are mostly dependent on the design of the building envelop variables (5). Building envelopes are in direct exposure to the external environment because of orientation and composition (6), and it is vital to identify the effects of their properties and optimise them in the early design phase. Moreover, they are responsible for more than half of the total heat gain in buildings (7). The building envelop design contains the building configuration, shell and material-related elements, including walls, floors, roofs and windows (8). These elements, through thermo-physical properties, represent the mechanisms of heat transfer or heat storage in the structure. Their geometries, construction materials and level of insulation are essential in this stage (9).

In this respect, U-value is the most important thermo-physical property that predominantly affects the rate of heat transfer in and out of a building and consequently the air conditioning or heating energy requirements (10). This value reflects how many Watts will flow in one hour through one square metre of an entire building section when the temperature variance between hot and cold side is 1° Celsius (11). Furthermore, the size and number of glazed windows in a building have a critical effect on both the heat and solar gains of a building. Since glazed areas have the highest heat gain per unit area, the majority of solar gains are passed through windows (12). The level of insulation also specifies the rate of heat conductivity and the infiltration of building components (11).

4.1.2   Building Layout

This category of variables mostly covers the design brief of the building, such as the internal sub-division, number of rooms or the number of spaces requiring heating and cooling, size of the building and building orientation. Generally, a larger building requires more energy to heat and cool than a smaller building because of the larger space to be heated or cooled (13). The question of whether a building needs less energy per unit volume or floor area, however, is more complex and still not completely resolved (14). Many researchers take the view that larger buildings need less energy per unit size because of their smaller surface area per unit size and thus lower heat gain per unit (15). Based on this perception, 'the larger a building, and the nearer to spherical in shape, the less are its energy needs because of the reduction in the ratio of surface area to volume' (16).

It can be stated that the architectural form of angular protrusions of buildings is an energy wasting form (16). However, compact buildings cost more to erect and have higher energy running costs than sprawling ones and the quality of 'compactness' in layout cannot be shown to be of paramount importance (17, 18). According to the literature, the maximum volume with minimum external wall perimeter cannot be regarded as the most EED (17). So, due to the need for mechanical systems in providing the interior comfort band, it may not even be a cheap design as well (19). As a whole, it does not seem feasible to generalise or quantify the complex implications that planning and layout of spaces have on air conditioning and lighting requirements (17). Building orientation also significantly affects the energy consumption of buildings. It changes the air conditioning or heating energy requirements in two respects (20):

1   Solar radiation and its heating effects on walls and rooms facing different directions
2   Ventilation effects associated with the relation between the directions of the prevailing winds and the orientation of the building

4.1.3   Occupant Behaviour

Among various factors influencing building energy consumption, occupant behaviour plays an essential role in energy performance, yet it is challenging to investigate due to complicated characteristics and unpredictable personal behaviour (21). The level of physical activity, clothing worn, duration of occupancy and age, size and background and behaviour of the occupant influence the cooling/heating requirements (22). For instance, a person wearing light clothes and doing light deskwork while seated will feel comfortable at 25° C, while a person wearing a light business suit may feel comfortable at 21° C (23). This 4-degree difference can mean a 100% difference in the cooling or heating energy requirement of a room. In addition, the attitude of the occupants toward energy use has significant consequences (24). The energy use is influenced by the aims and goals of the users, the penalties and benefits to the user of conserving energy and expectations of the user. Even a user's awareness of the relationship of their actions to the amount of air conditioning or heating energy used is another factor (25).

Building operation is also an important factor. A clear understanding of the building operation schedule is crucial to the overall accuracy of the energy estimation (24). This includes information regarding when building occupancy begins and ends (time of the week, seasonal variations), how many people are in the building and the schedule of internal equipment operations. The amount of energy used will be directly proportionate to the intensity of building occupancy (26). A building rented out for only half a year will obviously use half the energy of an equivalent building occupied throughout the year. Operating hours are, hence, the normalising factor that energy auditors must keep track of (27).

4.1.4   HVAC and Appliances

This category of variables includes the parameters related to heating, ventilation, air conditioning and electrical appliances within buildings (28). Energy management of this category is a primary concern, especially for commercial and office buildings since electricity has a considerable share in HVAC among all building services, installations and electric appliances (29). It is very complicated to model, simulate or optimise the energy consumption of these variables because, in addition to considering the thermal comfort and end user's behaviour, the efficiency and product quality of HVAC devices and appliances also play a major role in their energy consumption level (30). In fact, each appliance

Table 4.1 The Identified Variables from Literature

Variables	Physical Properties & Building Envelop	Building Layout	Occupant Behavior	HVAC	Reference
External Wall Material	✓				(6)
Internal Wall Material	✓				(32)
Roofing Material	✓				(6)
Type of Glazing in Windows	✓				(8)
Type of Glazing in Doors	✓				(12)
Level of Insulation	✓				(10)
External Shading	✓				(33)
Internal Shading	✓				(33)
Building Orientation		✓			(8)
Total Number of Rooms in the Housing Unit		✓			(16)
Metres Squared of Rooms Heated		✓			(34)
Metres Squared of Rooms Cooled		✓			(34)
Glazing to Wall Ratio		✓			(35)
Cross-Ventilation		✓			(14)
Ceiling Height		✓			(13)
Operation Schedule			✓		(24)
Clothes Type			✓		(22)
Electrical Appliance Usage			✓		(25)
Number of Occupants			✓		(26)
Type of Main Space Heating Equipment Used				✓	(34)
Type of Main Space Cooling Equipment Used				✓	(34)
HVAC Capacity				✓	(29)

has its different production configuration with different energy efficiency index. Thus, in order for a precise modelling, these interrelationships need to be recognised as well (31).

Table 4.1 indicates the identified variables from the literature along with their respected categories and source of references that were discussed in this section. There are 22 variables screened and identified from 16 referenced sources.

4.2   Delphi

Following an overview on building energy parameters, their categories and their summary in Table 4.1, and according to the problem to solution discourse in Chapter 1, the Delphi instrument comes to the fore in further identifying and prioritising building energy variables. This synthesised application of qualitative method (36) warrants that the research and findings remain linked to the literature and existing body of knowledge in the interim of creating new knowledge (37). The Delphi study is 'a qualitative method used to combine expert knowledge and opinion to arrive at an informed group consensus on a complex problem' (38). This method is not designed to collect random surveys, but it is an iterative discussion to obtain a consensus opinion from a relatively small sample of experts (39). In the literature, Delphi has been used in different contexts with different techniques; however, the basics of this facilitated method for complicated issues are similar (39):

- A possibility for respondents to convey their viewpoints on the topic
- A follow-up of feedback from respondents on their opinions
- Compiling the responses and analysing the group judgement
- A possibility for respondents to correct and/or modify their opinions according to the compiled responses
- An opportunity of consolidated agreement, anonymously

The complex problem targeted to this project is to complement the literature to identify and prioritise the variables that play key roles in energy consumption of residential buildings. Hence, the problem for Delphi study was broken down into a series of smaller inquiries addressing:

- Brainstorming with regard to the main variables contributing to the energy consumption and optimisation of residential buildings in the design stage
- Prioritisation of the identified key variables
- Confirmation of the process

4.2.1   Round 1

Round 1 was designed to provide a brainstorming session on building energy variables; in the first step, 35 variables were extracted from the qualitative responses through a quick textual analysis method and summarised in Table 4.2. The frequency of these parameters was counted, and it was observed that material referenced for 10 times is the most frequent variable while water fitting, water heating, structural design, shape and roof attic space parameters with only one time of reference are the least frequent variables.

Then, through a normative assessment method of the comparison of each variable within its peers, the variables were ordered from a range of 10 to 3 of the most frequent ones, resulting in 19 variables (Table 4.3). The list of 19 variables in this table is highly correlated with the outcome of the literature in Table 4.1. Generally, occupant behaviour is in the detailed level in the literature but is holistic from respondents' perspective. A possible reason could be in view of the respondents' consideration of the design stage as the scope of this research. Furthermore, lighting and daylighting factors are mentioned by

Table 4.2 35 Extracted Variables through a Quick Textual Analysis Method in Round 1

Variables	Reasons Mentioned by the Participants	Frequency
Space Heating and Cooling		1
Water Heating		1
Lighting		4
Material	Materials with different heat transfer coefficients are effective.	10
Shape		1
Morphology		4
Internal Walls Layout (Zones)		2
Window Glazing Type	Solar heat gain per orientation needed, sized and location.	9
Roofing Material Type		4
Ground Floor System		4
Structural and MEP Design		1
External Shading	Determines amount of sun/exposure to get sunlight in winter and block solar radiation in summer.	4
Orientation		8
Occupant Behaviour	Different people with different cultures have different routines that affect energy consumption.	8
Ceiling Height		5
Appliances and Equipment	Housewares with different energy labels result in different energy consumption.	2
Size of the Building		4
Size of the Opening		2
Insulation Type		9
Wall Thickness		4
Slab Insulation and Details		2
Air Distribution and Duct Design		2
Roof Attic Space		1
U-value of the Building Envelop	Indicate how much heat is lost through the walls, roof, floor, windows, doors.	1
Infiltration Rate		2
Window to Wall Ratio	Windows have a higher U-value than other thermal elements (e.g. the walls), so a greater source of potential heat loss.	8
Daylighting Factor	Indicating what proportion of natural light can replace artificial light in the building.	3
HVAC System		6
Colour of Façade	The colour of the façade, its colour and heat absorption or reflection factor can be an affective factor on the engineering design that should be discussed with the architect.	2
Internal Shading Type		3
Floor Area		2
Water Fittings		1
Solar Heat Gain Coefficient (SHGC)		2

Table 4.3 Normative Assessment Results

Variables	Frequency	Action
Material	10	Broken down into 'external wall material', 'internal wall material' and 'roofing material' as being generic.
Window Glazing	9	
Insulation Type	9	
Building Orientation	8	
Occupant Behaviour*	8	Removed as it is not in the scope of design stage and it goes to the operation stage.
Window to Wall Ratio	8	
HVAC System	6	Broken down into two variables of 'type of main space heating' and 'type of main space cooling'.
Ceiling Height	5	
Lighting	4	
Morphology*	4	Removed as being too subjective.
Roofing Material	4	
Building Size	4	Broken down into two variables of 'metres squared of rooms heated' and 'metred squared of rooms cooled'.
Ground Floor System	4	
Wall Thickness	4	
Internal Shading	3	
Daylighting Factor	3	
External Shading	3	
Door Glazing	3	

the respondents; however, these two items are ignored in the literature-based study. All in all, it can be stated that synthesising the outcomes of the literature and the brainstorming round of Delphi can provide more solid and established results in the next rounds.

In addition, the 19 derived variables were checked as to the implications for the applicability of variables for BIM, optimisation and design stage. In terms of BIM compatibility, very generic variables cannot be considered since the semantic values of very generic variables are not associated with the topological relationship within the BIM model. This idea is corroborated in the literature, where Kim and Anderson (40) indicated the lack of appreciation of too generic variables in the BIM environment. Therefore, a generic parameter of 'material' was divided into 'internal wall material', 'external wall material' and 'roofing material', falling to the predefined ranges of BIM families. In addition, variables with two-fold effects also need to be omitted because BIM has not been yet equipped with intelligent fuzzy rules to identify the root causes of two-fold parameters. Therefore, these variables cannot be optimised and should be removed. Ladenhauf, Battisti (41) confirmed this lack in their study on the role of computational geometries and intelligence in BIM interface.

Therefore, the variable of 'building size' was broken down into two variables of 'metres square of rooms heated' and 'metres squared of rooms cooled' (34). Additionally, in line with the two-fold notion, the HVAC system was further divided into two variables of the 'type of main space heating equipment used' and 'type of main space cooling equipment used' in order to consider both cooling and heating load applications (34) (Table 4.1). Checking the subjectivity and objectivity of the variables was another important task. Chong and Jun (42) indicated that too subjective variables can seriously hinder accurate energy estimation and simulation. Hence, the subjectivity and objectivity of the variables were controlled via their parametrisation and measurement capability in BIM. As a consequence, a too subjective variable, 'morphology', was omitted due to still not being parameterised and measured in BIM (41).

With respect to the design stage, the variables relevant to the occupancy profile of buildings cannot be incorporated as these variables fall to the operation phase of a building project lifecycle. This notion is further confirmed by the literature. It is widely believed that over buildings projects' lifecycle, design stage is the most promising phase in terms of EED as decisions here affect 60–70% of the lifecycle costs of construction and operation (43, 44). However, some variables seem to be more associated with the construction and operation phases. For example, the occupant behaviour category of variables refers to the people and their occupancy profile and behavioural use over the operation phase. In the design stage, it is difficult to predict the behaviour of occupants on the types of clothes worn, their schedule of building usage and their habits in using heating and cooling appliances (45).

In the final stage of the analysis of the first round, 19 variables obtained from the Round 1 of this Delphi study were depicted in Figure 4.1 to illustrate their schematic view. It can be seen that there is a wide range of variables that cover different aspects of buildings' energy parameters. External wall materials, internal wall materials, wall thickness, insulation type, roofing materials, ground floor system, window glazing types and door glazing types cover the physical properties and building envelop (4). There are parameters including ceiling height, metres squared of rooms heated, metres squared of rooms cooled, window to wall ratio and building orientation, which belong to the architectural design and building layout aspects of the influential parameters (17). Variables such as lighting, daylighting and external and internal shading address the required considerations in designing for energy efficient lighting and solar exposure (46). Finally, the parameters of cooling space equipment used and heating space equipment used apply the HVAC system design significance (28).

4.2.2   Round 2

This round was designated to focus on the prioritising aspect, and the experts were asked to prioritise the 19 variables derived from the first round. A five-point Likert scale questionnaire form was developed and the experts were asked to rate the variables ranging from 1 = least important, 2 = slightly important, 3 = important, 4 = very important to 5 = most important (47). Its rationale is that the dimension for measuring these variables should be unipolar, referring to different degrees of the same attribute, but not bipolar, referring to the presence of opposite attributes (48).

Daylighting Factor

External & Internal Shading

External Wall Material

Window Glazing Type

Window to Wall Ratio

Insulation

Lighting

Wall Thickness

Ceiling Height

Internal Wall Material

Doors Glazing Type

Building Orientation

Meter Square of Rooms Cooled

Meter Square of Rooms Heated

Roofing Material

Cooling Space Equip.

Heating Space Equip.

Ground Floor System

HVAC System

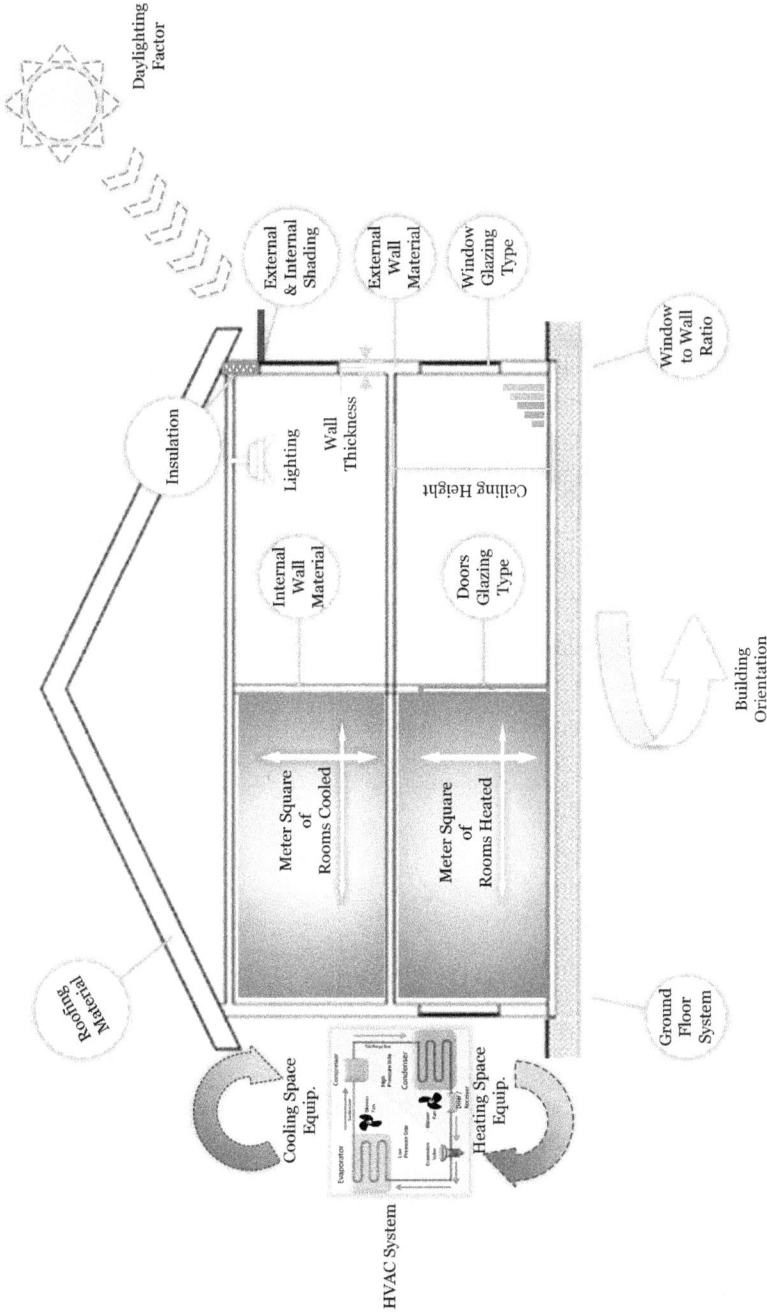

Figure 4.1 Schematic Illustration of the Variables Resulted from the First Round

A statistical analysis was conducted on the 18 questionnaires received in which the Likert point averages for the 19 resultant variables were calculated. A preliminary series of weighted variables was developed based on the mean ratings advocated by the 18 experts using the following equation:

$$L_{vi} = \frac{\sum_{i=1}^{n} P_{vi}}{n} , \qquad (4.1)$$

where L_{vi} stands for the Likert mean of each variable, P_{vi} is the point received for each variable (from 1 to 5) and n is the number of responses for each variable.

As a result of this round, the variables that received the L_{vi} equal to 3 or above were identified to be further analysed for the next round (47, 49). According to both Holt (47) and Perera, Rameezdeen (49), in the unipolar type of Likert scale, the point of 3 indicates the middle value in the importance scale. So, the analysis is based upon this middle range to determine the required level of significance for the parameters. Therefore, the parameters that receive 3 or above are regarded as significant and the parameters receiving the Likert point below 3 are determined as insignificant (47, 49). Table 4.4 shows the ranking of selected variables in the blue shaded and bold type, which amounts to 13 variables. It reveals that insulation, roofing material, external wall material, window glazing type and window to wall

Table 4.4 Results of Round 2 Questionnaires

The Variables	Likert Points Mean (L_{vi})	Rank
Insulation	**4.29**	**1**
Roofing Material	**4.17**	**2**
External Wall Material	**4.11**	**3**
Window Glazing Type	**4.00**	**4**
Window to Wall Ratio	**3.94**	**5**
Ceiling Height	**3.82**	**6**
Lighting	**3.58**	**7**
Metres Squared of Rooms Heated	**3.58**	**7**
Building Orientation	**3.52**	**8**
Metres Squared of Rooms Cooled	**3.52**	**8**
Type of Main Space Heating	**3.47**	**9**
Type of Main Space Cooling	**3.47**	**9**
Ground Floor System	**3.41**	**10**
External Shading	2.90	11
Wall Thickness	2.90	11
Door Glazing	2.90	11
Daylighting	2.50	12
Internal Wall Material	2.29	13
Internal Shading	2.29	13

ratio are the top five items. On the other hand, external shading, wall thickness, door glazing, daylighting, internal wall material and internal shading are the least important and were ignored for additional considerations.

The findings so far indicate that respondents emphasised the role of physical properties and building envelop in the energy consumption of residential buildings. The top three parameters of this ranking – insulation, roofing material and external wall material – receive the average Likert points of +4. These belong to the physical properties and building envelop category of variables. The reasoning behind this fact is that it is widely accepted that building envelop has direct exposure to the external environment. It channels a major path in thermal loss due to the heat transfer in winter and excessive solar heat gain area in summer (5). As a result, building envelop is responsible for more than half of the total heat gain in buildings (7). The outcomes also conform to the major trend in the energy optimisation-based studies in utilising the variables of three major categories of building envelop, building layout and HVAC systems in their simulation and analysis in the design stage (50–52). The significant variables identified through this Delphi cover a wide range of parameters involved in simulation and optimisation. Factors such as window glazing, window to wall ratio and types of main space heating and cooling loads are affiliated with building envelop, building layout and HVAC, respectively.

All in all, through the journey from the brainstorming step in Round 1 to the significant variable identification step via the Likert prioritisation in Round 2, around half of the items were selected by more than two-thirds of the experts. This fact implies that the majority of the participants agreed upon the significant variables in this round. For the sake of providing a measure of consistency, a statistical analysis was performed to compute Kendall's coefficient of concordance (W) (53) for the responses provided by the 18 experts. Kendall's W is a non-parametric test, running for the normalisation of Friedman statistic test, that can be used for assessing an agreement among participants. If the Kendall concordance coefficient equals 1, all the survey scorers have been unanimous and they rate the variables identically. On the contrary, if the test results in 0, it means that there is no overall trend of unanimity among the assessors and they rank completely differently (54).

For Kendall's W computation, S was first calculated from the row marginal sum of L_{vi} received by the variables:

$$S = \sum_{i=1}^{n} \left(L_{vi} - \overline{L_v} \right)^2,$$ (4.2)

where S is a sum of squares statistic over the sums of L_{vi}, $\overline{L_v}$ is the mean of the L_{vi} values and n is the number of variables. Following that, Kendall's W coefficient was obtained from the following formula:

$$W = \frac{12S}{m^2 \left(n^3 - n \right)},$$ (4.3)

in which m is the number of respondents and n is the number of variables; in this round, there are 18 respondents and 13 variables.

Table 4.5 The Concordance Measurement for Round 2

Variable	Likert Points Mean (L_{vi})	Mean Rank	S
Insulation	4.29	10.50	38.56
Roofing Material	4.17	10.09	35.04
External Wall Material	4.11	10.24	37.57
Window Glazing	4.00	9.38	28.94
Window to Wall Ratio	3.94	9.03	25.90
Ceiling Height	3.82	8.32	20.25
Lighting	3.58	7.56	15.84
Metres Squared of Rooms Heated	3.58	7.18	12.96
Building Orientation	3.52	7.44	15.36
Metres Squared of Rooms Cooled	3.52	7.18	13.39
Type of Main Space Heating	3.47	6.50	9.18
Type of Main Space Cooling	3.47	7.21	13.98
Ground Floor System	3.41	6.62	10.30
Number of Variables (n)			13
Number of Respondents (m)			18
Kendall's Coefficient of Concordance (W)			0.149
Level of Significance			10%

In this case, the Kendall's W test produced a score of 0.149 at 10% significance, which is higher than zero and means a trend of unanimity and consistency on the responses (54) (Table 4.5).

4.2.3   *Round 3*

The third round of this Delphi was devised for the confirmation purpose; the panel was provided with the results of the second round – those 13 significant variables along with their respected mean scores. The respondents were tasked with reconsidering their judgements in light of the mean scores received from their peers in the previous round.

Using Equation (4.1), Table 4.6 shows that the majority of the experts re-evaluated their rankings (Round 3 rankings in blue) in comparison with the previous round (Round 2 rankings in grey). External wall material was raised from the third to the top level, while insulation and roofing material were lowered to the second and third rates, respectively. This implies that the building envelop has still caught the first consideration. Likewise, the respondents directed their particular attention to the external wall material because of its recognised effect on the heat transfer of building fabric (11). Furthermore, lighting dropped from seventh to nineth, and type of main space heating rose by 4 rankings. Interestingly, the twin variables of metres squared of rooms heated and cooled became convergent and took eighth and ninth places. It should be noted that the Likert points mean (L_{vi}) of all variables kept the score above 3, which confirms an appropriate congruence among the results (47, 49). In line with

Table 4.6 Results of Round 3 Questionnaires

The Variables	Likert Points Mean (L_{vt})	Rank Round 3	Rank Round 2
External Wall Material	4.46	1	3
Insulation	4.38	2	1
Roofing Material	4.15	3	2
Window Glazing	4.00	4	4
Building Orientation	3.84	5	8
Window to Wall Ratio	3.76	6	5
Ceiling Height	3.61	7	6
Type of Main Space Heating	3.53	8	9
Type of Main Space Cooling	3.53	8	9
Metres Squared of Rooms Heated	3.53	8	7
Metres Squared of Rooms Cooled	3.30	9	8
Lighting	3.30	9	7
Ground Floor System	3.23	10	10

Table 4.7 The Concordance Measurement for Round 3

Variable	Likert Points Mean (L_{vt})	Mean Rank	S
External Wall Material	4.46	12.00	56.82
Insulation	4.38	11.19	46.31
Roofing Material	4.15	10.42	39.26
Window Glazing	4.00	9.69	32.37
Building Orientation	3.84	9.00	26.56
Window to Wall Ratio	3.76	8.62	23.53
Ceiling Height	3.61	7.92	18.52
Type of Main Space Heating	3.53	7.23	13.62
Type of Main Space Cooling	3.53	7.35	14.52
Metres Squared of Rooms Heated	3.53	7.35	14.52
Metres Squared of Rooms Cooled	3.30	6.23	8.53
Lighting	3.30	6.23	8.53
Ground Floor System	3.23	5.62	5.70
Number of Variables (n)			13
Number of Respondents (m)			15
Kendall's Coefficient of Concordance (W)			0.267
Level of Significance			10%

the previous round, the consistency of the survey was again calculated via Kendall's coefficient of concordance (W) using Equations (4.2) and (4.3). It was revealed that the consistency was remarkably enhanced in the third round, where it attained a number of 0.267 at 10% significance level. This points out 80% improvement in reliability and consistency as compared to the second round (Table 4.7).

4.3   Summary

The current BIM does not sufficiently support decision-making procedures to indicate a roadmap of energy optimisation of the significant variables of the design
through the perspectives of BIM-compatible variables. The identified variables
in this chapter could be the key asset for energy optimisation of buildings in the
design stage if they are encapsulated through a unified approach in the BIM design
process. To this end, the parametric definition inherited in the BIM technology
should be focused. These variables then must be parametrised via the BIM development underlying approaches: AI such as machine learning and data transformation algorithms in order to be approachable in the BIM environment via securing
the geometrical, topological and semantic associations (Figure 4.2). The integrated
and parametric platform of BIM allows for the implementation of 'what if scenarios' so that the optimisation techniques and algorithms are run on these variables to find the most optimum values and to minimise the energy consumption of
buildings in the design stage.

This chapter was intended to conduct data collection and analysis to identify
the significant parameters of EED, scoped to the design stage and to be compat-
ible with BIM according to the literature and experts' responses. The sequential
exploratory research method was run by synthesising the literature and three-round
Delphi. This procedure ended up with a final matrix of 13 key parameters along
with the rankings and mean scores, fulfilling a reliable concordance analysis and
comprising major categories of physical properties and building envelop, building
layout and HVAC.

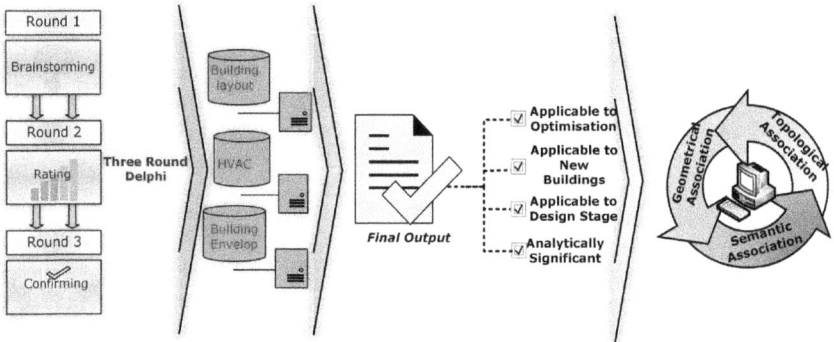

Figure 4.2 Graphical Diagram of the Steps Leading to the Output

References

1 Radhi H. A systematic methodology for optimising the energy performance of buildings in Bahrain. Energy and Buildings. 2008;40(7):1297–303.

2 Clarke JA. Energy simulation in building design: Routledge; 2001.

3 Park J, Park J, Kim J, Kim J. Building information modelling based energy performance assessment system: An assessment of the energy performance index in Korea. Construction Innovation: Information, Process, Management. 2012;12(3):335–54.

4 Nemirovski G, Sicilia A, Galán F, Massetti M, Madrazo L, editors. Ontological representation of knowledge related to building energy-efficiency. Proceedings of the Sixth International Conference on Advances in Semantic Processing; 2012.

5 Koo C, Park S, Hong T, Park HS. An estimation model for the heating and cooling demand of a residential building with a different envelope design using the finite element method. Applied Energy. 2014;115:205–15.

6 Wong SL, Wan KK, Li DH, Lam JC. Impact of climate change on residential building envelope cooling loads in subtropical climates. Energy and Buildings. 2010;42(11):2098–103.

7 Utama A, Gheewala SH. Indonesian residential high rise buildings: A life cycle energy assessment. Energy and Buildings. 2009;41(11):1263–8.

8 Banihashemi S, Golizadeh H, Hosseini MR, Shakouri M. Climatic, parametric and non-parametric analysis of energy performance of double-glazed windows in different climates. International Journal of Sustainable Built Environment. 2015;4(2):307–22.

9 Bouchlaghem N. Optimising the design of building envelopes for thermal performance. Automation in Construction. 2000;10(1):101–12.

10 Ekici BB, Aksoy UT. Prediction of building energy needs in early stage of design by using ANFIS. Expert Systems with Applications. 2011;38(5):5352–8.

11 El Hjouji ME, Khaldoun A, editors. Thermal auditing of buildings: Essential step towards designing energy efficient houses. Renewable and Sustainable Energy Conference (IRSEC), 2013 International; March 7–9, 2013.

12 Urbikain M, Sala J. Analysis of different models to estimate energy savings related to windows in residential buildings. Energy and Buildings. 2009;41(6):687–95.

13 Gong X, Akashi Y, Sumiyoshi D. Optimization of passive design measures for residential buildings in different Chinese areas. Building and Environment. 2012;58(0):46–57.

14 Anastaselos D, Oxizidis S, Manoudis A, Papadopoulos AM. Environmental performance of energy systems of residential buildings: Toward sustainable communities. Sustainable Cities and Society. 2016;20:96–108.

15 Michalek J, Choudhary R, Papalambros P. Architectural layout design optimization. Engineering Optimization. 2002;34(5):461–84.

16 Standard M. MS 1525: 2007: Code of practice on energy efficiency and use of renewable energy for non-residential buildings. Malaysian Standard; 2007.

17 Cao J, Metzmacher H, O'Donnell J, Frisch J, Bazjanac V, Kobbelt L, et al. Facade geometry generation from low-resolution aerial photographs for building energy modeling. Building and Environment. 2017;123:601–24.

18 Gros A, Bozonnet E, Inard C, Musy M. Simulation tools to assess microclimate and building energy: A case study on the design of a new district. Energy and Buildings. 2016;114:112–22.

19 Bucking S, Zmeureanu R, Athienitis A. A methodology for identifying the influence of design variations on building energy performance. Journal of Building Performance Simulation. 2014;7(6):411–26.

20 Kibert CJ. Sustainable construction: Green building design and delivery: Wiley; 2012.

21 Yu ZJ, Haghighat F, Fung BC, Morofsky E, Yoshino H. A methodology for identifying and improving occupant behavior in residential buildings. Energy. 2011;36(11):6596–608.

22 Soebarto V, Bennetts H. Thermal comfort and occupant responses during summer in a low to middle income housing development in South Australia. Building and Environment. 2014;75:19–29.

23 Böer KW. Advances in solar energy: An annual review of research and development: Springer Science & Business Media; 2012.

24 Delzendeh E, Wu S, Lee A, Zhou Y. The impact of occupants' behaviours on building energy analysis: A research review. Renewable and Sustainable Energy Reviews. 2017;80:1061–71.

25 Masoso O, Grobler LJ. The dark side of occupants' behaviour on building energy use. Energy and Buildings. 2010;42(2):173–7.

26 Amasyali K, El-Gohary NM. Energy-related values and satisfaction levels of residential and office building occupants. Building and Environment. 2016;95:251–63.

27 Yu Z, Fung BC, Haghighat F, Yoshino H, Morofsky E. A systematic procedure to study the influence of occupant behavior on building energy consumption. Energy and Buildings. 2011;43(6):1409–17.

28 Jung DK, Lee DH, Shin JH, Song BH, Park SH. Optimization of energy consumption using BIM-based building energy performance analysis. Applied Mechanics and Materials. 2013;281:649–52.

29 Bichiou Y, Krarti M. Optimization of envelope and HVAC systems selection for residential buildings. Energy and Buildings. 2011;43(12):3373–82.

30 Pérez-Lombard L, Ortiz J, González R, Maestre IR. A review of benchmarking, rating and labelling concepts within the framework of building energy certification schemes. Energy and Buildings. 2009;41(3):272–8.

31 Capone A, Barros M, Hrasnica H, Tompros S, editors. A new architecture for reduction of energy consumption of home appliances. Towards eEnvironment, European Conference of the Czech Presidency of the Council of the EU; 2009.

32 Tuhus-Dubrow D, Krarti M. Genetic-algorithm based approach to optimize building envelope design for residential buildings. Building and Environment. 2010;45(7):1574–81.

33 Manzan M, Pinto F, editors. Genetic optimization of external shading devices. Proceedings of 11th International IBPSA Conference, Glasgow, Scotland; 2009.

34 Korolija I, Marjanovic-Halburd L, Zhang Y, Hanby VI. Influence of building parameters and HVAC systems coupling on building energy performance. Energy and Buildings. 2011;43(6):1247–53.

35 Waldron D, Jones PJ, Lannon SC, Bassett T, Iorwerth HM. Embodied energy and operational energy: Case studies comparing different urban layouts. 2013.

36 Merriam SB. Qualitative research: A guide to design and implementation: Jossey-Bass; 2014. Available from: http://deakin.eblib.com.au/patron/FullRecord.aspx?p=1662771.

37 Bazeley P. Qualitative data analysis: Practical strategies. SAGE; 2013.

38 Donohoe HM, Needham RD. Moving best practice forward: Delphi characteristics, advantages, potential problems, and solutions. International Journal of Tourism Research. 2009;11(5):415–37.

39 Ameyaw EE, Hu Y, Shan M, Chan AP, Le Y. Application of Delphi method in construction engineering and management research: A quantitative perspective. Journal of Civil Engineering and Management. 2016;22(8):991–1000.

40 Kim H, Anderson K. Energy modeling system using building information modeling open standards. Journal of Computing in Civil Engineering. 2012;27(3):203–11.

41 Ladenhauf D, Battisti K, Berndt R, Eggeling E, Fellner DW, Gratzl-Michlmair M, et al. Computational geometry in the context of building information modeling. Energy and Buildings. 2016;115:78–84.

42 Chong I-G, Jun C-H. Performance of some variable selection methods when multicollinearity is present. Chemometrics and Intelligent Laboratory Systems. 2005;78(1):103–12.

43 Stumpf A, Kim H, Jenicek E, editors. Early design energy analysis using BIMs (building information models). Building a Sustainable Future: Proceedings of the 2009 Construction Research Congress; 2009.

44 Kim H, Stumpf A, Kim W. Analysis of an energy efficient building design through data mining approach. Automation in Construction. 2011;20(1):37–43.

45 Yan D, O'Brien W, Hong T, Feng X, Gunay HB, Tahmasebi F, et al. Occupant behavior modeling for building performance simulation: Current state and future challenges. Energy and Buildings. 2015;107:264–78.

46 Harish V, Kumar A. A review on modeling and simulation of building energy systems. Renewable and Sustainable Energy Reviews. 2016;56:1272–92.

47 Holt GD. Asking questions, analysing answers: Relative importance revisited. Construction Innovation. 2014;14(1):2–16.

48 Schwarz N. Cognition and communication: Judgmental biases, research methods, and the logic of conversation: Psychology Press; 2014.

49 Perera B, Rameezdeen R, Chileshe N, Hosseini MR. Enhancing the effectiveness of risk management practices in Sri Lankan road construction projects: A Delphi approach. International Journal of Construction Management. 2014;14(1):1–14.

50 Nguyen A-T, Reiter S, Rigo P. A review on simulation-based optimization methods applied to building performance analysis. Applied Energy. 2014;113:1043–58.

51 Machairas V, Tsangrassoulis A, Axarli K. Algorithms for optimization of building design: A review. Renewable and Sustainable Energy Reviews. 2014;31:101–12.

52 Foucquier A, Robert S, Suard F, Stéphan L, Jay A. State of the art in building modelling and energy performances prediction: A review. Renewable and Sustainable Energy Reviews. 2013;23:272–88.

53 Kendall MG, Smith BB. The problem of m rankings. The Annals of Mathematical Statistics. 1939;10(3):275–87.

54 Corder GW, Foreman DI. Nonparametric statistics for non-statisticians: A step-by-step approach: John Wiley & Sons; 2009.

5 AI Algorithms Development

5.1 Dataset Generation

The simulation method was selected in this stage to collect reliable and veri-fied data for the prioritised variables with an emphasis on covering the whole range of available values. These data are collected via a building energy per-formance simulation. As outlined in one of the seminal studies by Karlsson and Roos (1), there are four main methods to develop an energy simulation model:

1. Comparison based on the physical properties
2. Using empirical coefficients based on the energy balance of components for different climatic conditions and building orientation
3. Incorporating simple building properties to distinguish between different building types
4. Performing a full-scale simulation including climatic data, building layout, building envelop and physical properties

Method 4 provides very accurate results (provided the simulation model is cor-rect) as compared to the others since it includes a wide range of parameters (2). Nevertheless, it also requires experienced users, a lot of input data, which are not generally available, and huge computation for simulation (2). Given the advantage of the accurate and comprehensive simulation in AI-based research (3), this study used the fourth method for generating a comprehensive dataset of energy relevant inputs and outputs in a broad context via considering the differ-ent climatic data.

However, in order to overcome the above-mentioned drawbacks of the fourth method, this study incorporated an approach in the size reduction of the full facto-rial energy simulation by using a metaheuristic method. The metaheuristic method assists in obviating the dataset development from numerous runs of simulations and limits the number of input data to those that were determined significant in the Delphi study (4). It is based on the fact that changing the values of variables in each run has direct impacts on the heating and cooling loads of buildings. This

DOI: 10.1201/9781003207658-5

fact consequently affects the ultimate energy consumption of the simulated case (2). To set the scope of the simulation, the following criteria were also taken into consideration:

- The simulated case should represent a typical residential building type in the considered climatic areas inclusive of the specifications complied with the well-known and international standard of ASHRAE (5) in its building envelope.
- The simulation model should encompass the main types of climatic conditions of temperate, tropical, hot-arid and cold (6).

Considering all these elements, a five-story building consisting of four units per floor along with the parking lot in the ground floor was selected for simulation. Each level area is 320 m², summing up to a total area of 1600 m². The building was modelled in Rhino 5 and parameterised in Grasshopper software[1] (elaborated in the next section). Figure 5.1 shows the typical layout of the floors and the perspective of the building. Table 5.1 lists the components and construction used in the simulation model, which conforms to the variables resulting from the Delphi and their specification sets based on ASHRAE 90.1–2007.

For clarification of the simulation process and the construction specifications used, the model, layout and materials were constructed based on ASHRAE 90.1–2007. The identified variables from Chapter 4 were corresponded to their required construction specifications in the standard and their respective climatic zones (Table 5.1). This procedure enabled the researcher to generate a baseline model according to the identified variables, their required specifications from ASHRAE standard and the climatic zones of temperate, tropical, cold and hot-arid. For building envelop and physical properties, available wall, roof, floor, insulation and glazing types for the selected climatic zones were extracted and incorporated into the baseline model (see Table 5.1). Building orientation was also considered for its main eight different degrees. Steam or hot water system, heat pump and built-in room heater were selected for different types of main space heating and central systems and window to wall units were considered for the different types of main space cooling systems (5). Metres squared of rooms heated and cooled were set for 10 and 15 metres, and the ceiling height was considered for a range of eight 3–3.5 m, which are applicable for each room in a residential unit (7). The window to wall ratio of 20–40% and lighting range of 1–40 lux were set as recommended by (5) (Table 5.1).

Among various factors influencing residential building energy consumption, occupant behaviour plays an essential role but is difficult to analytically investigate due to its complicated characteristics (8). It was discussed in Chapter 4 that occupancy variables could not be considered for estimation and optimisation as the focus of this book is on the design stage. However, since the Rhino and Grasshopper package utilises the EnergyPlus engine in its energy calculation procedure (9), some basic occupancy parameters such as user profile should be selected by

Figure 5.1 3D Model and Layout

Table 5.1 Model Parameters Specifications based on ASHRAE 90.1–2007

Components	Specifications	U-Value (W/m².K)
	ExtWall Mass Climatezone 1	0.7
	ExtWall Mass Climatezone Alt-Res 1	0.98
	ExtWall Metal Climatezone 1–6	0.7
	ExtWall Mass Climatezone 2	0.98
	ExtWall Mass Climatezone Alt-Res 2	0.77
	ExtWall Mass Climatezone 3	0.77
Wall Type	ExtWall Mass Climatezone Alt-Res 3	0.64
	ExtWall Mass Climatezone 6	0.48
	ExtWall Mass Climatezone Alt-Res 6–7	0.42
	ExtWall Steel Frame Climatezone 1–2	0.78
	ExtWall Wood Frame Climatezone 1–4	0.54
	ExtWall Wood Frame Climatezone 6–7	0.3
Roof Type	ExtRoof Iead Climatezone 1	0.37
	ExtRoof Iead Climatezone 2–8	0.28
	ExtRoof Metal Climatezone 1–7	0.38
Insulation Type	ExtWall Insulation Layers 1–8	NA
	ExtRoof Insulation Layers 1–8	NA
Floor Type	Floor Climatezone 1–8	0.19
	Floor Climatezone 2–7	0.15
Glazing Types	Double-Glazed Window	0.29
	Triple-Glazed Window	0.28
Building Orientation	0, 45, 90, 135, 180, 225, 315	NA
Type of Main Space Heating	Steam or Hot Water System	NA
	Heat Pump Built-in Room Heater	
Type of Main Space Cooling	Central System Window/Wall Units	NA
Metres Squared of Rooms Heated/Cooled	10 m² 15 m²	NA
Ceiling Height	3 m 3.5 m	NA
Window/Wall Ratio	20–40%	NA
Lighting	1–40 Lux Range	NA

default to enable the software to proceed with the simulation processes. Hence, in order to facilitate this procedure, it was taken into account that the heating or cooling system in the building becomes active when the inside temperature goes higher or lower than the predefined comfort band.

For energy simulation in the EnergyPlus plugin, a building calculation program was set to low-rise apartment; the kitchen, bedroom, bathroom and dining room in each unit were defined as separate zones with their own thermal properties. This approach enables the thermal engine to precisely quantify adjacencies and inter-zonal connections (10). The thermostat was set between 18 and 26 °C (11) to provide thermal comfort for occupants and activate HVAC devices below or above this range. Table 5.2 indicates the user profile of the zones for the units of the building. From Figure 5.1, it can be seen that some units have three bedrooms and some units are single bedroom.

Using different climatic conditions in building energy performance simulation assists in generating the wide range of datasets and enhances the generalisability of the output and the framework developed for energy optimisation purposes (12). Therefore, considering the dominant climatic zones introduced by Kottek, Grieser (6), four cities of Sydney, Moscow, Kuala Lumpur and Phoenix were chosen as representatives of temperate, cold, tropical and hot-arid climates, respectively. Such a wide range of climatic situations allows for developing more generalisable simulation dataset. Table 5.3 indicates the geographical coordination of these cities along with the dominant thermal systems required for each city (13). The relevant EnergyPlus Weather format data were downloaded from the EnergyPlus database (13) and embedded in the simulation software indicating the climate zones of the four cities. These climatic data are in accordance with the typical metrological year (TMY) (14) concept in which datasets of hourly values of solar radiation and meteorological elements are recorded in a 12-year period of time and averaged for a 1-year timespan. Table 5.4 shows the monthly mean values of climatic data for the selected cities, including temperature, humidity, wind speed and solar radiation.

Table 5.2 User Profile

Zone	Area (m^2)	Volume (m^3)	Occupancy	Activity	Comfort Band
Living Room	19	57	4	Sedentary	18–26 °C
Kitchen	7	21	4	Cooking	NA
Bathroom	4	12	1	Sedentary	NA
Bedroom	10	30	2	Sedentary	18–26 °C

Table 5.3 Cities Chosen for Simulation

City	Latitude	Longitude	Climate	Dominant Thermal System
Sydney	33.86°S	151.2°E	Temperate	Cooling and Heating
Moscow	55.75°N	37.61°E	Cold	Heating
Kuala Lumpur	3.13°N	101.68°E	Tropical	Cooling
Phoenix	33.44°N	112.07 W	Hot-arid	Cooling and Heating

Source: Adapted from WMO (15)

Table 5.4 Climatic Data of the Selected Cities

City	Month	Temperature (°C)	Humidity (%)	Wind Speed (k/h)	Average Daily Solar Radiation (MJ/m²)
Sydney	Jan	22.3	62	20	21.4
	Feb	22.3	64	25	19
	Mar	21.2	62	22	16.4
	Apr	18.5	59	20	12.9
	May	15.5	57	8	10
	Jun	13.1	57	14	9.1
	Jul	12.4	51	9	9.8
	Aug	13.4	49	16	13
	Sep	15.5	51	12	16.7
	Oct	17.8	56	22	19.9
	Nov	19.5	58	22	21.3
	Dec	21.33	59	22	22.6
Moscow	Jan	−8	90	17	2.3
	Feb	−7	85	14	5.3
	Mar	−2	75	14	9.9
	Apr	6	70	14	14.2
	May	13	70	14	18.1
	Jun	18	75	14	18.8
	Jul	19	75	14	18.5
	Aug	17	80	14	15.2
	Sep	12	85	16	9.4
	Oct	6	85	16	5.3
	Nov	−1	90	16	2.7
	Dec	−6	90	17	1.5
Kuala Lumpur	Jan	27	80	9	15.6
	Feb	27	80	12	18.4
	Mar	27	80	6	18.9
	Apr	28	85	6	19.3
	May	28	80	9	18.2
	Jun	27	80	9	17.7
	Jul	27	80	12	17.6
	Aug	27	80	12	17.9
	Sep	27	80	9	18.1
	Oct	27	85	8	17.4
	Nov	27	85	8	15.3
	Dec	27	85	6	13.7
Phoenix	Jan	13.6	45.6	8.2	11.7
	Feb	15.4	43.9	9	14.8
	Mar	18.4	37.7	10.3	19.9
	Apr	22.6	26.2	11.3	25.3
	May	27.8	20.6	11.3	27.9
	Jun	32.7	17.9	10.8	28.1
	Jul	34.9	29.1	11.3	25.1
	Aug	34.2	34.6	10.8	22.4
	Sep	31.3	31.1	10	20.9
	Oct	24.8	31.4	8.9	16.7
	Nov	17.8	36.7	8	13.1
	Dec	13	46.2	7.7	10.8

Source: Adapted from WMO (15)

5.2   Data Size Reduction

5.2.1   *An Overview*

The full factorial inclusion and all possible combinations of 13 parameters together and their values and ranges in the energy simulation (see Table 5.1) requires a huge number of simulation runs. It may require more than 6 billion simulations,[2] which is completely infeasible to conduct. As a solution, data size reduction techniques are commonly used in the whole building energy performance simulation (16). Not doing so in the studies that focus on the annual aspects of energy and consist of high temporal resolution (seconds to hours) restricts the spatial resolution to a rough zonal discretisation (17). In recent years, different approaches have been used to minimise the number of simulations required and to simplify the building energy simulation procedures to glean precise but concise results (18–20).

According to Eriksson, Johansson (21), Design of Experiment is one of the most commonly applied approaches that can be used to design any information-oriented experiment, especially where the variation and its observation are of high importance. It is a very efficient branch of parametric tests for evaluating the effects and possible interactions of several factors through its factorial design tool (22). The two-level full factorial design in Design of Experiment relies on the approximation of the model by the polynomial expansion of 2^k possible combinations of the factors multiplied by the specified range of the levels of values (23). For instance, the Design of Experiment factorial design of a batch of three parameters with two levels for each factor would generate eight runs of simulations. However, this method is applicable for continuous parameters and cannot handle categorical variables due to its numerical nature (24). Therefore, in this book, a novel technique was applied via exploiting the parametric nature of Grasshopper software in incorporating generative design principles in the energy simulation development, as discussed next.

5.2.2   *Metaheuristic-Parametric Approach in Data Size Reduction*

Parametric modelling is a computational method, capable of delivering both generative and analytical model and streamlines a dramatic shift from modelling a designed object to the design logic (25). This method utilises the computational attributes in setting the design principles to provide a platform of design exploration and variations. In fact, different degrees of AI are applied to the computational specifications, such as rules, constraints, parametric dependencies and heuristic and metaheuristic structures to encode data. It acts as a generator in order to yield a parametric-generative model (26). The procedure of parametric-generative design constitutes three major elements of (27):

1   Start conditions and parameters (input)
2   A generative mechanism (rules, algorithms etc.)
3   The act of generation of the variants (output)

Each generative process starts with the inputs to establish the initial parameters, which are then transformed through a generative mechanism toward the initial population of design. This mechanism is a finite set of instructions, rules and/or algorithms to fulfil a specific purpose in a finite number of steps. Upon the generation of variants, multiple design variants, a benchmarking or a selection procedure should be determined in the identification of the best variant and final output (27). It is widely recognised that there is not a single and definite solution; rather, an iterative divergence/convergence process is required to deliver the most comprehensive range of possibilities. This iteration helps explore, analyse and identify the best design option with regard to the desirable criteria (28).

In line with the underlying concept of generative design, as mentioned earlier, the Rhino and Grasshopper software package was used as an integrated computer design tool with algorithmic method. Parametric modelling tools can simplify the widest possible range of concepts for design exploration by allowing the automatic generation of a group of alternative design solutions (9). Rhino is a 3D modelling software that authorises the designer to link the layout to its underlying parameters through a plugin called Grasshopper. Grasshopper is also regarded as one of the most suitable parametric modelling platforms embedded in Rhino for developing the design variant algorithms in light of its powerful parametric programming capabilities (29). It is a graphical algorithm editor, tightly integrated with Rhino 3D modelling tools, which feature an advanced user interface. The major interface of the algorithm development in Grasshopper applies the node-based editor in which data is processed from a component by connecting wires. These wires always connect an output grip to an input grip where data can either be defined as a constant or imported as a variant parameter. Therefore, in this book, the following steps were taken into consideration:

1   The variables and their respected values and ranges were parameterised and linked to the building 3D model (see Table 5.1).
2   The selection procedure of construction details, including wall, insulation, roof and floor for four cities, was hinged on their respected climate zones, according to ASHRAE standard (5). That is, an item index moves among the possible options of component categories and locks on the appropriate item that matches with the relevant climate zone. The number of moves depends on the size reduction rules set by heuristic principles (Figure 5.2).
3   The 'holistic cross-reference' command was used to set the back-and-forth iterative connections among all parameters and their values. This procedure imitates the full factorial simulation in the traditional approach. Figure 5.3 exemplifies this procedure, in which two set elements of A, B, C and X, Y, Z are iterated equally nine times (3*3) and then crossed into each other, creating nine new members of AX, AY, AZ, BX, BY, BZ, CX, CY and CZ. (The cross-referencing of these elements imitates the cross-referencing of the whole parameters used in the model simulation.)

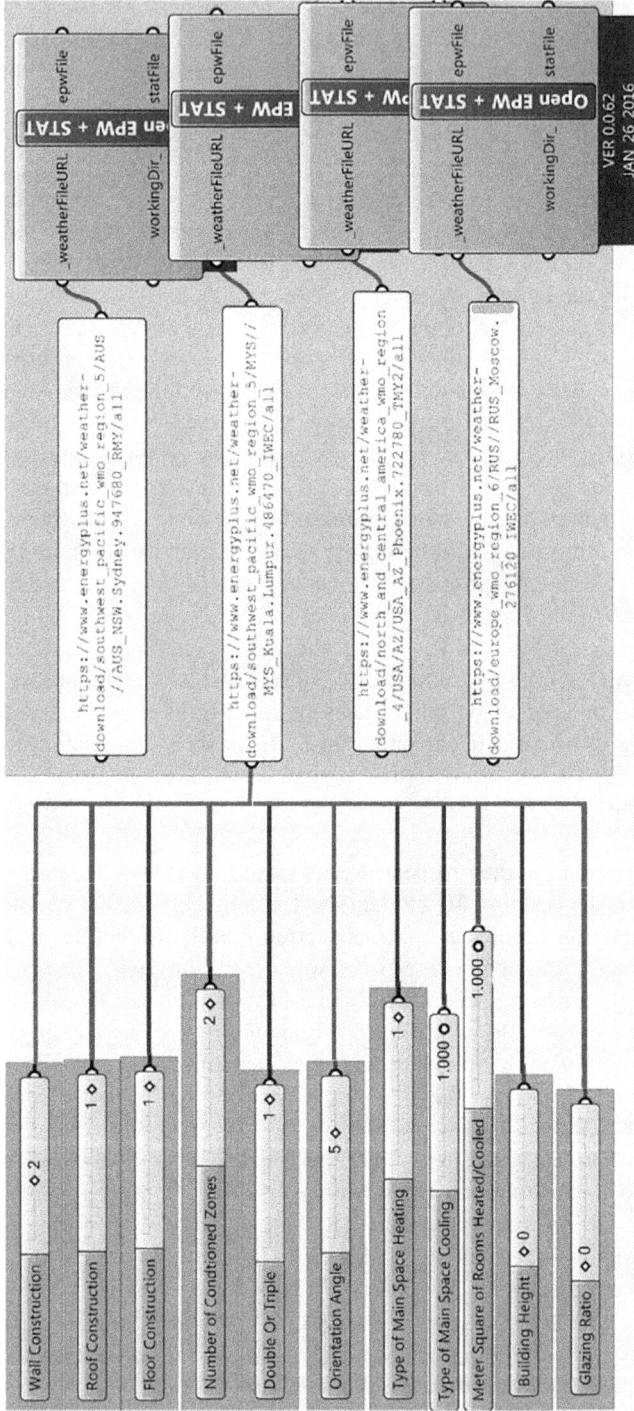

Figure 5.2 Parametric Setting of the Variables

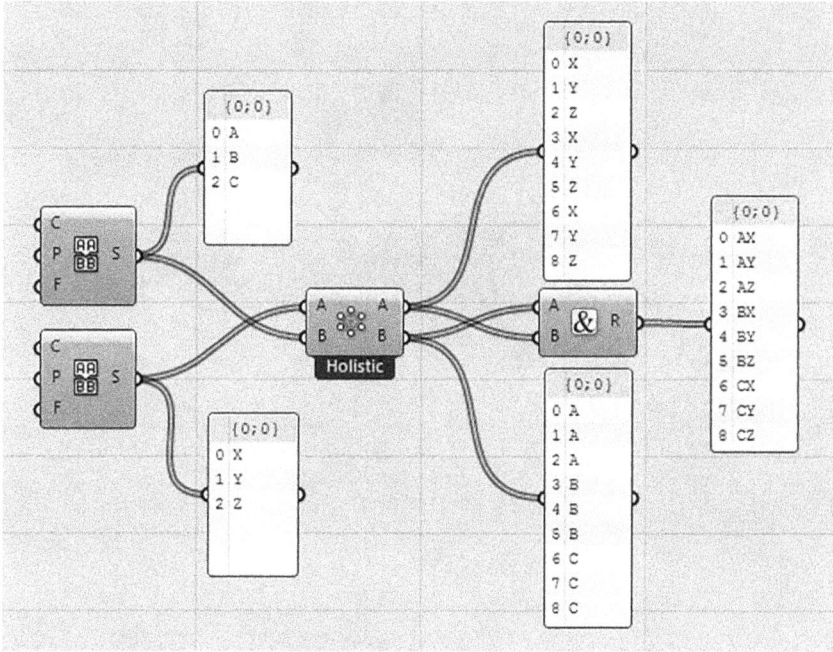

Figure 5.3 Holistic Cross-Reference

4   Metaheuristic principles of an evolutionary solver function were set upon the parameters in order to implement the virtual data size reduction and limit the full factorial simulation by using these rules:

- Max stagnant was empirically fixed on five iterations for the initial population of 20 generations of genomes (13 parameters). This rule means that if the fitness of five consecutive iterations comprising 20 generations of 13 genomes is kept fixed, the evolutionary solver function (energy simulation) is terminated.
- To control the distance of genomes in each generation, the inbreeding rate was determined on 75% and the digital variant was maintained on 5%. This rule indicates how the relative offset of genomes (parameters) could be guided in distributing the mutation and crossing over of parameters for securing an appropriate functional handling in dataset generations. Figure 5.4 illustrates this; next to the population generation arising from the previous steps, the genomes (simulation variables) are mutated and crossed over. This covers the whole range of data so that the presented metaheuristic rules pick the most optimums up.

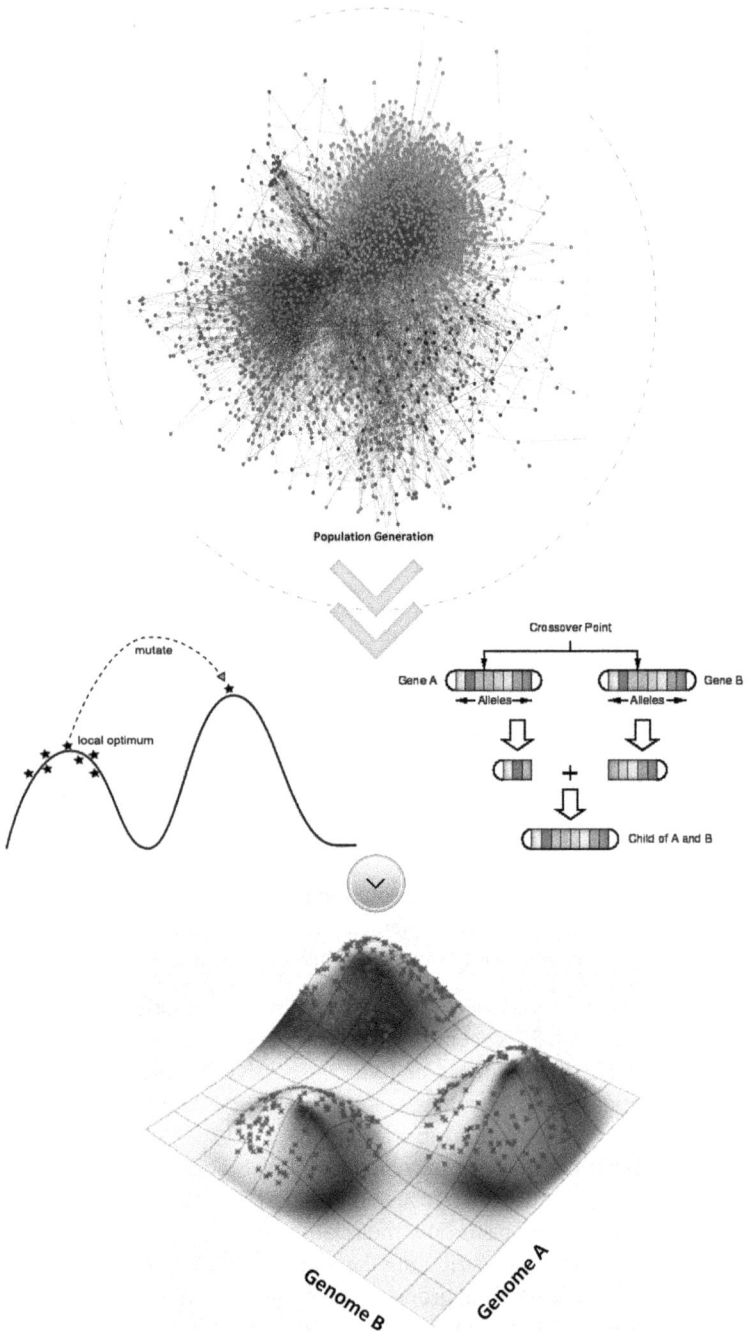

Figure 5.4 Conceptual Diagram of Heuristic Data Size Reduction
Source: Adapted from Talbi (30) and Gendreau and Potvin (31)

5.3 Data Interpretation Approach

The procedures taken for the data generation resulted in 4,435 runs of simulation, including 13 inputs (the variables) leading to outputs (annual energy consumption) consisting of 1053, 1138, 1114 and 1130 data for the four cities of Sydney, Phoenix, Kuala Lumpur and Moscow, respectively. The differences in the numbers of generated data arise from the metaheuristic data size reduction in light of their initial population generation and the number of iterations toward the convergence of the evolutionary solver function. For this reason, the solutions found are dependent on the set of random populations generated (30). To analyse the possible impacts of such randomness and obtain an overview about the generated dataset, descriptive statistics were performed for the whole dataset using Statistical Package for Social Science (SPSS) software on the measures of central tendency and dispersion (32).

Different statistical measurements of minimum, maximum, range, mean, median, mode and standard deviation were computed to specify the probability distribution and the dispersion of the whole dataset (Table 5.5). This information demonstrates the status of the generated data vis-à-vis the normal distribution state. Table 5.5 indicates that by considering both continuous and categorical data, the generated output for all cities covers a wide range of distribution. This fact is of advantage for optimisation purposes because it enables more precise optimisation (33).

Matlab software was also applied to plot the observations with respect to the outputs, annual energy loads (Figure 5.5). It should be noted that the number of observations in this figure refers to each batch of inputs with 13 variables. This figure illustrates a relatively steady state from the first to the mid-observation range,

Table 5.5 Descriptive Statistics of the Developed Dataset

Parameters	Range/ Number	Minimum	Maximum	Mean	Std. Deviation	Median
Wall Type*	6.00	NA	NA	NA	NA	NA
Insulation Type*	7.00	NA	NA	NA	NA	NA
Roofing Material*	2.00	NA	NA	NA	NA	NA
Window Glazing Type*	2.00	NA	NA	NA	NA	NA
Ground Floor System*	2.00	NA	NA	NA	NA	NA
Type of Main Space heating*	3.00	NA	NA	NA	NA	NA
Type of Main Space Cooling*	2.00	NA	NA	NA	NA	NA
Building Orientation	315.00	0	315	102.98	87.02	90.00
Window to Wall Ratio	0.20	0.2	0.4	0.25	0.08	0.20
Ceiling Height (m)	0.50	3.00	3.50	3.14	0.22	3.00
Metres Squared of Rooms Heated	5.00	10.00	15.00	11.77	2.39	10.00
Metres Squared of Rooms Cooled	5.00	10.00	15.00	12.39	2.49	10.00
Lighting (Lux)	39.00	0.00	40.00	2.39	2.58	2.00
Energy Load (kWh)	10535.58	309.88	10845.45	2725.16	2015.30	2208.66

* Categorical parameters

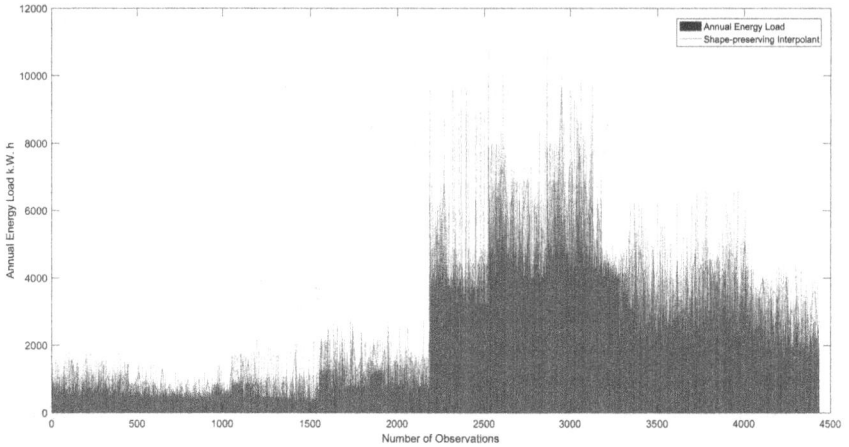

Figure 5.5 Annual Energy Load vs. Number of Observations[3]

where the annual energy loads of Sydney and Phoenix are depicted up to 2000 kWh. This monotonous trend is in view of the climatic conditions of these two cities (see Table 5.4), which refuse the excessive amount of energy consumption for keeping the comfort band. For the other two cities, however, Kuala Lumpur and Moscow, a dramatic shift occurs from the middle of the observations to the end and the annual energy loads are made at least double, reaching to more than 10,000 kWh on some occasions. Such an increase is quite normal as Kuala Lumpur and Moscow are in the extreme climatic regions: tropical and cold, respectively.

5.4   AI Development

5.4.1   Introduction

The application of metaheuristic size reduction, coupled with the data interpretation and interpolation approach, ended up with the interpolated dataset. The well-fitted dataset is now prepared for use in developing the optimisation algorithm. But it should be noted that each optimisation algorithm requires a specific objective function as its fundamental unit to model the parameters and minimise the values. Developing the most appropriate objective function is a critical task as much as the reliable dataset generation, particularly where both types of qualitative and quantitative data are involved with the optimisation solver (33). It was revealed in Chapter 2 that a wide range of AI algorithms was applied for prediction, classification or optimisation of buildings' energy consumption.

However, there is a salient gap in the investigation of hybrid objective function development for energy optimisation problems, including qualitative and quantitative datasets in their constructs. It is widely believed that two types of machine learning algorithms, namely prediction and classification, are run for continuous and discrete parameters of building energy consumption, respectively (34). In some cases, transformation techniques are also utilised for converting the continuous and discrete variables into each other, but some shortcomings, such as losing integrity or randomness, may arise (35). Therefore, in this stage, two well-known algorithms of ANN and DT were developed separately and integrally upon the datasets in order to find the best solution for energy optimisation functions. Chapter 2 presented the discussion on the common predictive methods in building energy performance, including CDA, SVM and ANN. From these discussions and the pros and cons of different methods, ANN was chosen for this study in view of its fast-computing capability, great level of accuracy, handling large amounts of data and its wide application in building energy prediction (34, 36, 37).

Furthermore, among the different classification algorithms, DT is one of the most effective supervised classifiers that works very well with both continuous and categorical data (38). Comparing DT with SVM as one of the other powerful classifiers: DT takes less time in the computation time and runs effectively on the non-linear data (39). In addition, Caruana and Niculescu (40) analysed ten different classification algorithms on eleven different datasets and compared the results on eight different performance metrics. Results indicated the superior performance of DT as compared to others such as SVM, random forest and Naïve Bayes.

5.4.2 Artificial Neural Network

Artificial neural network (ANN) is a mathematical or computational model that tries to simulate the structure or functional aspects of biological neural networks and presents deep learning opportunity (36). One of the applications of ANN in the engineering field is to predict the outcome of non-linear statistical problems, and it is usually utilised to model complex relationships between inputs and outputs or to find patterns in datasets (41). The thermal equations used to analyse and calculate energy loads are complex, making ANN a good platform for this purpose (2). In this form, the network was presented with datasets obtained from simulations and the values of inputs were fed into each neuron or node. The weights were then iteratively adjusted through learning algorithms until a suitable output was produced (34). A suitable output, in this case, suitable predicted annual energy load, is the one that is as close as to the simulation results as possible.

One of the most popular and efficient network structures for an ANN model is the multilayer perceptron (MLP). MLP consists of identical interconnected neurons that are organised in layers (36). These layers are also connected to which outputs of a layer act as the inputs of subsequent layers. Data flow starts from the input layer and ends in the output layer (42). Through this journey, data pass through one or multiple hidden layers that recode or provide a representation for the inputs. In this study, due to the large number of variables and existing non-linearity among them, MLP network was used to model significant relationships between the inputs and the output and predict the data-driven energy performance.

5.4.2.1   ANN Model Configuration and Performance Analysis

This study constructed a three-layer ANN model of feed-forward type with one output neuron. There are 13 neurons in the input layer for the 13 input variables in the model and one neuron in the output layer. In ANN modelling, the data are divided into three groups of training, testing and validating. The best configuration of ANN models usually depends on some elements, such as the number of neurons in the hidden layer, the type of learning algorithm and the training–testing proportion of data (37). The identification of the best configuration is a trial-and-error process that requires different experiments to find the best architecture and optimal performance (36).

Therefore, this process commenced with the best architecture identification for this ANN model. When using multilayer neural networks for solving a problem, the number of neurons in the hidden layers is one of the most important issues. It is known that an insufficient number of neurons in the hidden layers leads to the inability of neural networks to solve the problem (36). On the other hand, too many neurons lead to over-fitting and decreasing of network generalisation capabilities due to increasing the freedom of the network more than is required (41).

Although the selection of architecture for ANN comes down to trial and error, the best number of neurons for the hidden layers could be experimented on with a few heuristic rules (43):

- The number of hidden layer neurons is equal to the number of neurons in the input layer, or
- The number of hidden layer neurons is equal to two times the number of input layer neurons plus one, or
- The number of hidden layer neurons is equal to the number of input layer neurons plus number of output layer neurons, or
- The number of hidden layer neurons is equal to the sum of the number of input layer neurons and the number of output layer neurons divided by two.

Therefore, the four different numbers of 13, 7, 14 and 27 neurons for one hidden layer, considering the default types of Matlab software package for learning algorithms (Levenberg-Marquardt), transfer functions of hyperbolic tangent and sigmoid and 1000 iterations were tested. Consequently, the number of 7 neurons was found to be the most optimum number for the hidden layer neurons. Best performance of the model was measured based on the error produced by the ANN model, which in this case, mean square error (MSE) was used as a performance indicator. MSE can give a quantitative indication of the model error in terms of a dimensional quantity (44). MSE equal to zero indicates a perfect match between the observed and predicted values and is calculated by the following equation:

$$MSE = \frac{\sum_{i=1}^{N}\left(E_p - E_a\right)^2}{N},$$

(5.1)

where E_a is the actual energy value, E_p is the predicted energy value and N is total number of datasets (41). The model including 7, 13, 14 and 27 hidden neurons resulted in, respectively, 0.048, 0.061, 0.058 and 0.073 errors. Thus, the optimum number of neurons was found to be 7. The conceptual structure of this ANN model is visualised in Figure 5.6.

In the ANN conceptual architecture, the information flow starts at the input layer, ending in the output layer, and this happens through the hidden layer (41). Subsequent to model architecture identification, the best configuration selection was followed with the optimum training algorithm. Hence, six different types of training algorithms with a bias toward the back propagation category as the most efficient for MLP (45) were tested. This test was run via 7 hidden neurons, the default transfer functions of hyperbolic tangent and sigmoid and 1000 iterations and their performance errors were calculated (Table 5.6). Back propagation is a method that feeds back the size of the error into the calculation for the weight changes (46). According to Table 5.7, the Levenberg-Marquardt back propagation algorithm was found to have the least algorithm error (0.0029) vis-à-vis the benchmark error (0.005) asserted by Flores (41). So, the Levenberg-Marquardt back propagation algorithm was used as a method to fit the weights during the learning process starting at the output layer and through the input layer (Table 5.7).

To find the optimum percentage of dataset[4] to be trained, tested and validated, test 1 with 60% training – 20% testing – 20% validating; test 2 with 70% training – 15%

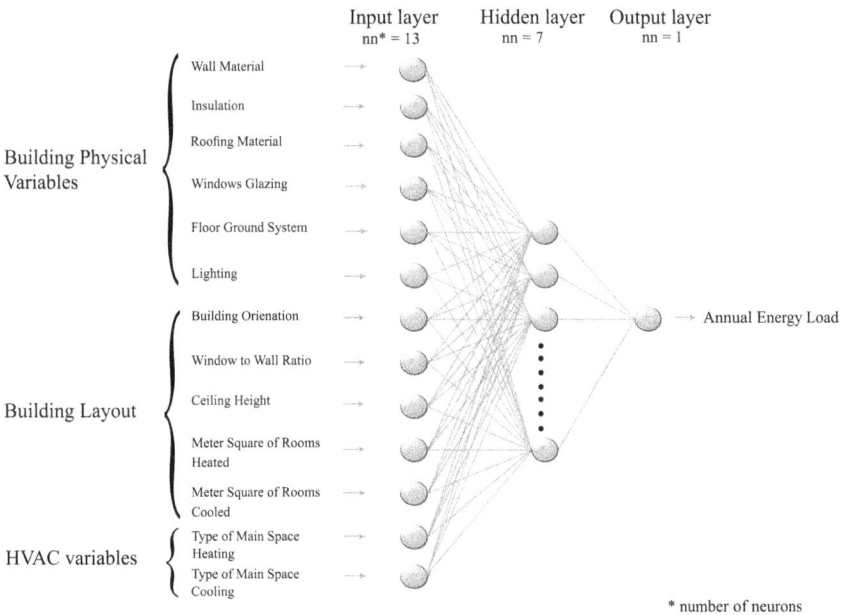

Figure 5.6 Conceptual Architecture of the Developed ANN

Table 5.6 ANN Training Algorithms Applied

Number	Function ID	Algorithm Name	Seminal References
1	Trainbr	Bayesian regularisation back propagation	(47)
2	Trainscg	Scaled conjugate gradient back propagation	(48)
3	Trainlm	Levenberg-Marquardt backpropagation	(49)
4	Trainoss	One-step secant back propagation	(50)
5	Trainrp	Resilient back propagation	(51)
6	Traingd	Gradient descent back propagation	(52)

Table 5.7 Different Training Algorithms Performance

Number	Function ID	Benchmark Error	Algorithm Error	Number of Iterations
1	Trainbr	0.005	0.0043	1000
2	Trainscg	0.005	0.0062	1000
3	**Trainlm**	**0.005**	**0.0029**	**1000**
4	Trainoss	0.005	0.0038	1000
5	Trainrp	0.005	0.0057	1000
6	Traingd	0.005	0.0047	1000

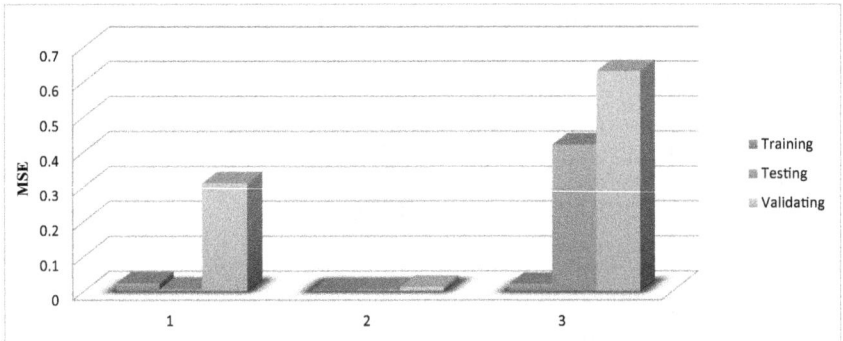

Figure 5.7 Different Training, Testing and Validating Percentage Performances

testing – 15% validating; and test 3 with 80% training – 10% testing – 10% validating were performed, as recommended by Shahidehpour, Yamin and Li (43). The observations were randomly used for training, testing and validating since the random observations in ANN design and development are imperative to avoid biases and evaluate ANN performance more robustly (36). Based on the results, the lowest MSE for training, testing and validating, with values of 0.011 and 0.008, were achieved by test 2 when 70%, 15% and 15% of the dataset were used for training, testing and validating, respectively (Figure 5.7). Furthermore, it can be stated that this batch of percentage is deemed appropriate in terms of providing a sufficient number of cases for a proper procedure of training, testing and validating. This is according to the seminal reviews of similar studies in the literature (4, 53–55).

5.4.2.2 Final ANN Model

Given the identification of the best architecture and configurations for ANN model so far, the final run of ANN with 13, 7 and 1 neurons in input, hidden and output layers was started including 70% training, 15% testing and 15% validating. Hyperbolic tangent function was the activation function chosen for the input layer, sigmoid transfer function was applied between the hidden layer and the output layer and the Levenberg-Marquardt back propagation algorithm was set as the learning algorithm. The program was then instructed to run for 1000 iterations as maintained by Demuth, Beale (36) and Shahidehpour, Yamin and Li (43), and the error for each run of iteration was measured. In this model, 1000 iterations were found to be adequate for the optimal training process.

The iteration should be terminated when no obvious change and/or improvement is observed. Hence, in order to avoid overtraining, it was intended that training stop when the error remains unchanged for six iterations (56). Overtraining has additional popular meaning in ANN structure design. If too many hidden layer neurons are used, ANN is trained to keep too many details of the training data, and so it performs much worse in the testing data (36). Figure 5.8 illustrates that

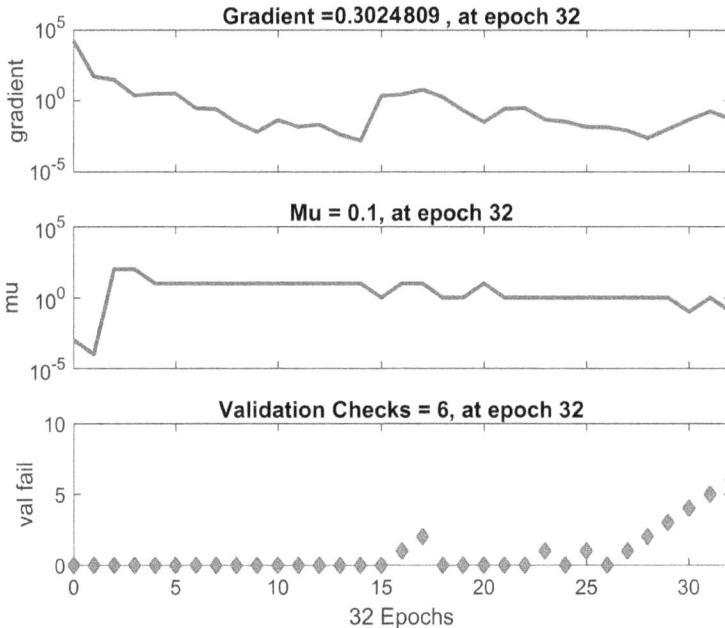

Figure 5.8 ANN Training State

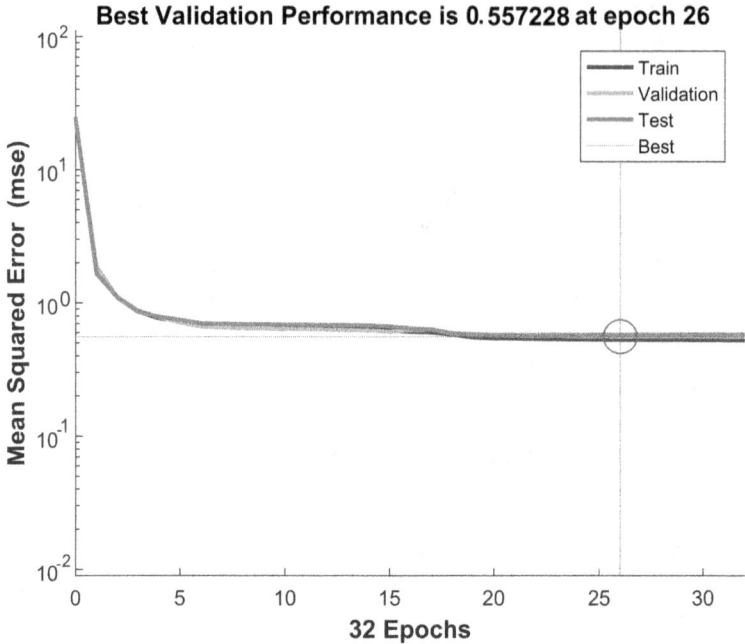

Figure 5.9 Best Validation Performance

the minimum gradient and the MSE for training data was recorded at the 32nd epoch. This amount of epoch and the respected error were deemed acceptable in comparison with the literature (41, 57).

As the training stops after six consecutive increases in the validation error, the best performance was taken from the 26th epoch with the lowest validation error (Figure 5.9). Furthermore, graphically demonstrating the model performance, the correlation coefficient graph (R) indicates how much close predicted energy loads fit to the actual loads. Closer R value to 1 shows that predicted values are closer to the actual results (58). Figure 5.10 illustrates that the overall R value of the developed ANN was calculated at 0.81186. This number presents a strong performance as an objective function in the optimisation problem in comparison with the statistical perfection of 1. This achievement is also remarkable as compared to the similar studies in the literature (10, 53, 55).

5.4.3   Decision Tree

5.4.3.1   An Overview

In Section 5.4.1, it was discussed that because of including both continuous and categorical data in the current dataset, it is of priority in running two major

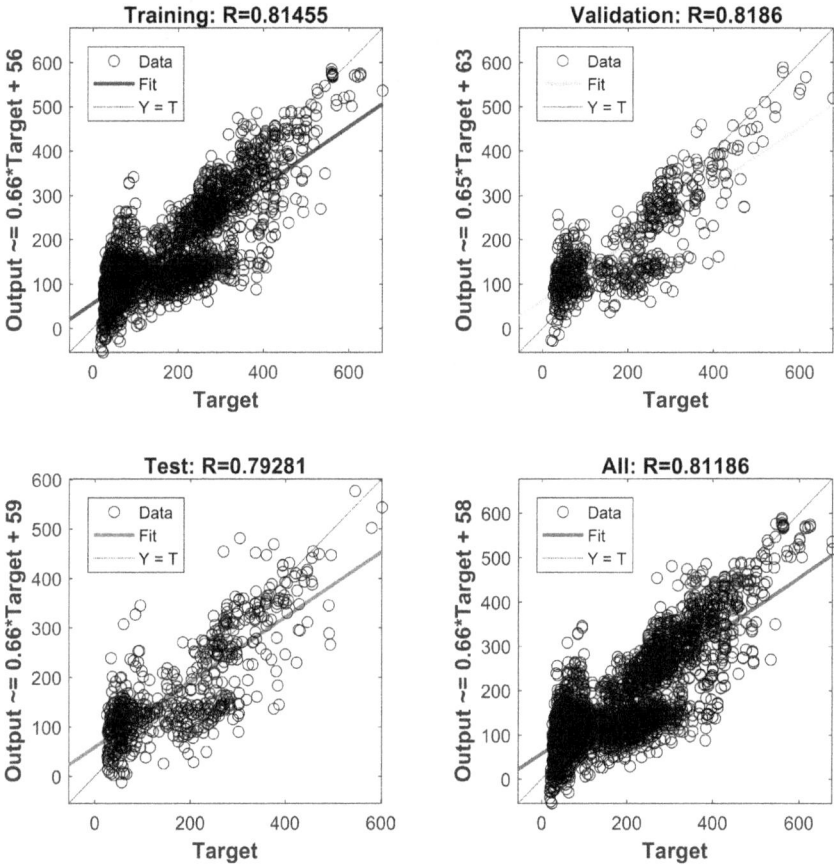

Figure 5.10 Regression Test of Final ANN Model

categories of machine learning applications on prediction and classification algorithms, separately and integrally. Thus, next to the application of ANN prediction
on the whole dataset in Section 5.4.2, this section applies the classification algorithm DT on the whole dataset. It is further followed by Section 5.4.4 to apply the
combination of these two algorithms on the same dataset. This investigation may
facilitate knowledge for finding a more proper solution with better performance
and lower error.

DT is one of the most applied types of machine learning algorithms in classification problems (38, 59). The reputation of this algorithm is largely attached to its
interpretability and accuracy in delivering classification models with understandable structure, which generates useful information on the corresponding domain.
In addition, DT is capable of processing both numerical and categorical parameters
(58). However, it should be considered that this method is more appropriate and

accurate in handling categorical parameters rather than numerical data (60). There are different types of DT algorithms, including simple tree, medium tree, complex tree and bagged tree, which follow similar fundamental principles but are of different degrees of complexity in combining trees. It applies a flowchart-like tree structure to separate the dataset into different predetermined categories for presenting the interpretation, categorisation and generalisation on data (61).

The tree-based logical model of the algorithm denotes the process of target (dependent variable) prediction and classification via the values of a batch of independent parameters. A tree algorithm consists of three basic components of root node, branch and leaf (62):

- A root node evaluates a specific attribute.
- Branches are assigned to the node according to the available values for each attribute.
- A leaf shows a class, and when an item attains a leaf, the leaf's class will be designated to the item.

Root node and branches indicate a binary split test on a specific attribute, whereas leaf node presents the result of the classification and holds the categorical label.

5.4.3.2   DT Model Configuration and Performance Analysis

Generally, DT generation follows three steps, as suggested by Muniyandi, Rajeswari and Rajaram (63):

1   Learning the data
2   Attribute selection
3   Optimum DT architectural configuration

With regard to the first step, learning the data, the energy dataset (as a result of dataset generation; Section 5.1) should be prepared for use as training data. Hence, the continuous parameters were first assigned with the classes based on the integer numbers and the output was classified considering the classes set to the continuous parameters. The target variable for this DT is the annual energy load, with four potential states classified as low, medium, high and excessive energy consumption as recommended by Yu, Haghighat (60). This means that the algorithm is going to be fully developed in order to identify the appropriate output ranges for subjective classifications. Similar to ANN, in DT's learning process, data should be divided into two subsets of training and testing in which a majority is usually employed for training and the rest for testing. However, unlike ANN, the proportion of training and testing in DT depends on the algorithm to learn, and this element is not controlled by the user. The underlying DT generation algorithm is ID3 (64), which was further developed and extended to C4.5 algorithm (65) and its flexibility and applicability in dealing with different data types were improved. It adopts the training dataset as the input and delivers the decision structure.

C4.5 algorithm is a top-down and greedy search through the space of possible branches with no backtracking. It iteratively splits a partition by choosing a split attribute to well separate the target class values. This process is initiated with grouping the training data into a single partition. In each iteration, a predictor attribute is selected by the algorithm that can best split the target class values in that partition. Following with the predictor attribute selection; C4.5 algorithm separates the partition into child partitions in a way that each child partition consists of the same value of the selected attribute. Iteratively splitting the partitions, the learning algorithm terminates when one of the following circumstances is met (60):

1   All data within a partition features the same target class value. As a result, the class label of the leaf node is the target class value.
2   Further splitting a partition is not feasible due to running out of all remained predictors. So, the leaf node is labelled with the majority target class values.
3   No data are available for a particular value of a predictor variable. Thus, a leaf node is created with the majority class value in the parent partition.

In building a DT, the second step is to choose an attribute (as a node) and then divide it (branching). Since C4.5 partitions the data into subsets that should contain instances with similar values, this homogeneity should be calculated in a structured manner (38). Therefore, the concept of entropy as maintained by Shannon and Weaver (66) is usually used to assess the homogeneity of the data samples in DT. If the sample is quietly homogenous, the entropy would be 0, and if the sample is an equally separated, it will have the entropy of 1.

$$E(D) = E(P_1, P_2, \ldots, P_n) = \sum_{i=1}^{n} -P_i \log_2 P_i, \qquad (5.2)$$

where E and D denote the entropy and the energy dataset, n indicates the number of values that the final classes can take (four values of low, medium, high and excessive) and P_i is the partition of the dataset (D) ranging from one to four. The information-gain is computed based on the decrease in entropy after a dataset is split on an attribute. Building a DT completely falls to finding an attribute that delivers the highest information-gain or the most homogenous branches. Thus far, subsequent to calculating entropy for different branches using Equation (6.4), they are accumulated to obtain the total entropy for the splits and the information-gain could then be estimated via the following equation:

$$IG(D, A) = E(D) - \sum_{v \in Values(A)} \frac{|D_v|}{|D|} E(D_v), \qquad (5.3)$$

where IG and A are information-gain and attribute, D_v is the subset of D (dataset) with $A = v$ and *Values* (A) is the set of all possible values of A. The attribute with the largest information-gain is selected as the split attribute for each tree node, but this metric is biased toward the attributes consisting of a large number of domain values. Normalisation of the information-gain by a split information value is a solution to minimise such a bias, which could be considered in a C4.5 algorithm structure (67):

$$Gain\ Ratio\ (D,A) = \frac{IG(D,A)}{Split\ Info\ (D,A)}, \tag{5.4}$$

where

$$SplitInfo\ (D,A) = \sum_{i=1}^{n} \frac{|D_i|}{|D|} \log_2 \frac{|D_i|}{|D|}. \tag{5.5}$$

Ultimately, a branch with entropy of zero is set as a leaf node and branches with more than zero require further splitting. C4.5 is run recursively on the non-leaf branches until no critical observation can be found on information-gain or gain ratio in further splitting. In this case, it means that all data are classified (68).

Similar to ANN model development, the configuration of DT algorithm also falls to the trial-and-error process (61). Therefore, it is reasonable to test different tree structures and find the best performance among them. Through training the data and attribute selection, the Classification Learner Toolbox of the Matlab software package was then used on developing the various types of DT and the generated outcomes were recorded. All possible types of DT, including simple tree, medium tree, complex tree and bagged tree (ensemble trees) algorithms (65), were developed using the energy dataset in which the architecture of simple tree (Figure 5.11), medium tree (Figure 5.12) and complex tree (Figure 5.13) were

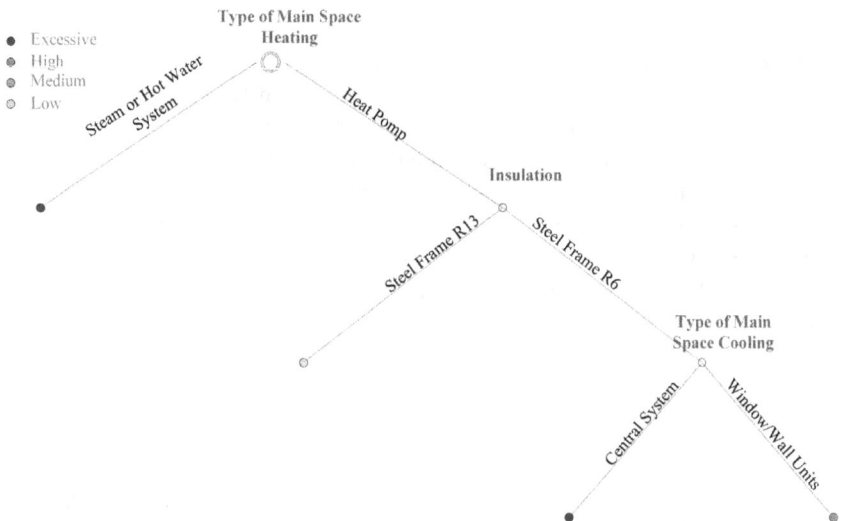

Figure 5.11 Simple Tree

visualised. It is worth mentioning that the architecture of bagged tree is not possible to demonstrate due to the massive interactions of more than 100 complex trees with each other.

The root node and starting point of the classification algorithms for all types of DTs is the type of main space heating parameter, determined by Equation 5.3. Using Equations 5.4 and 5.5 in the developed trees, decision rules can be utilised starting from the root node and falling through branches to the leaf nodes. For instance, in Figure 5.11, it can be seen that the simple tree, via choosing a steam or hot water system as the type of main space heating, is classified as the excessive energy consumption category. This tree could reach the medium level of energy consumption by selecting the steel frame R6 and window to wall unit as the insulation and main space cooling type, respectively.

Such a decision becomes more informed when medium tree is applied and more parameters come to the fore. As depicted in Figure 5.12, in addition to the type of main space heating parameter, which is the root node of the algorithm, four layers of branches consisting of seven unique leaf nodes (parameters) are added. Consequently, extracting diverse decision rules allows for more precise classification. For example, heat pump for main space heating in combination with steel frame R13 (insulation), window to wall ratio of 20% and steel frame 1–4 (wall) could lead to low energy consumption of residential buildings.

The more complicated tree is developed, the more conscious decisions will be made (68). Figure 5.13 illustrates a complex tree that comprises nine layers of branches and all 13 parameters of the energy dataset. As mentioned earlier, the interpretative capability of DT distinguishes this algorithm from its counterparts. Valuable information can be obtained from this DT model, which provides a blueprint for energy consumption pattern and its optimisation objective functions (38). For instance, different variables are automatically chosen as the precursors for classifying the energy consumption level. These variables are selected to split the

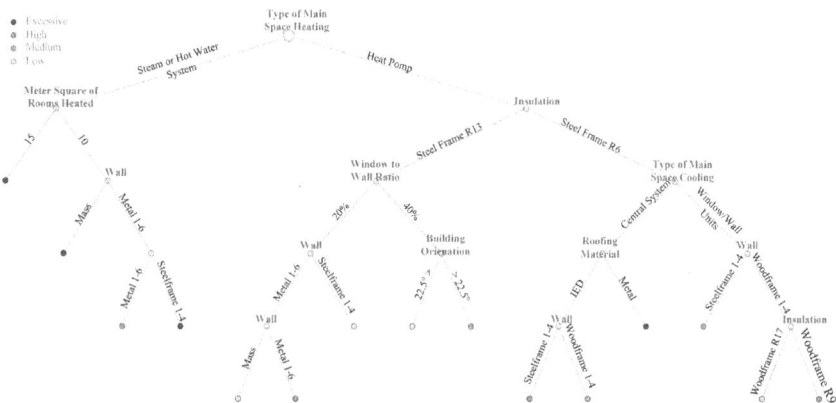

Figure 5.12 Medium Tree

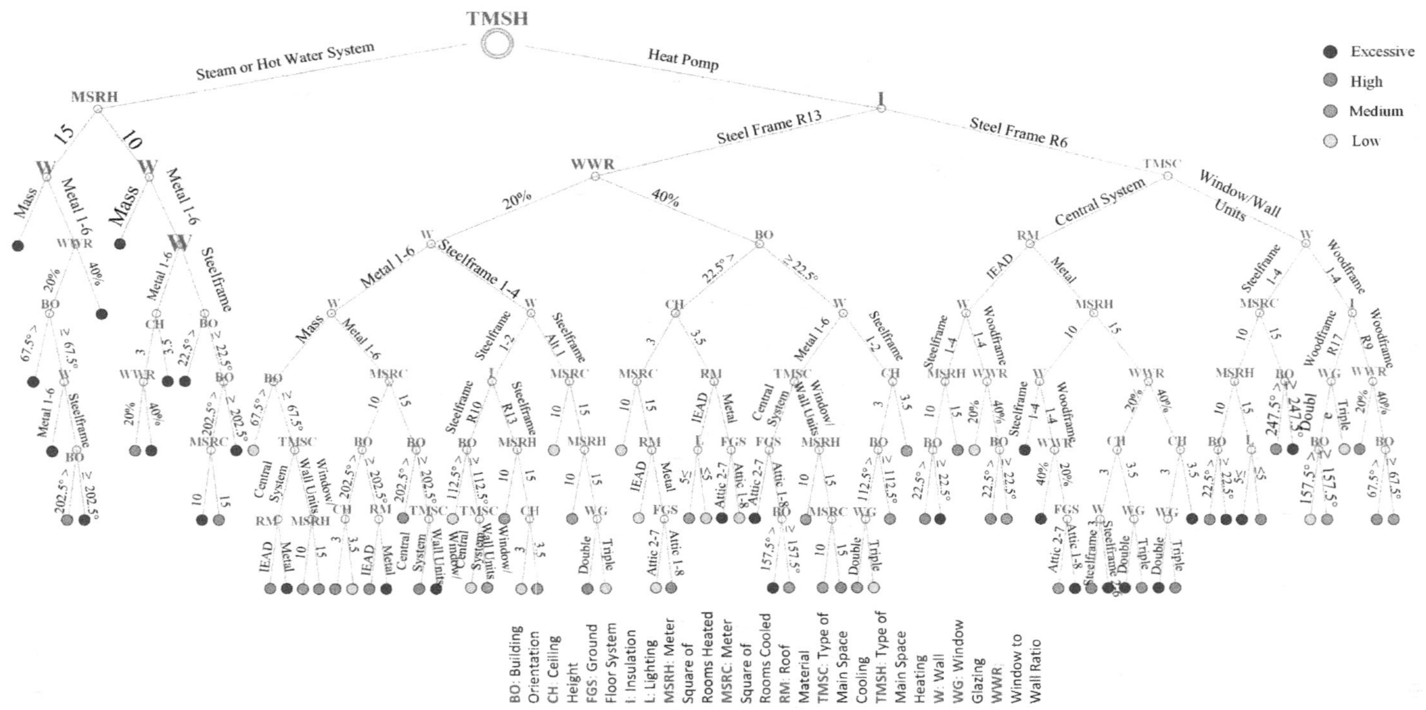

Figure 5.13 Complex Tree

The fourth type of DT, bagged tree, is an ensemble learning method using bootstrap aggregation, which bags a weak learner such as a complex DT on the dataset. It generates many bootstrap replicas of the dataset and grows decision trees on each replica (69). Bagging works by training learners on multiple-complex DTs on the dataset. In addition, every tree in the ensemble can randomly select predictors for decision splits that improve immensely the accuracy of the bagged trees. The optimum number of complex DTs that should be bagged and aggregated is dependent on the classification error, generated through testing and cross-validation processes (68).

nodes of DT, and their distance from the root node implies their significance and the number of records influenced. Particularly, the visualised complex tree covers a wide range of attributes for all parameters and traversing across the branches and leaf nodes result in different decision rules. Hence, by exploring the tree nodes, the level of strength of the variables and their ranks, which specify the building energy demand profile, can be extracted.

Hence, 200 complex trees were aggregated and bagged, and the produced errors in testing and cross-validating states are displayed in Figure 5.14. This figure shows that the performance of the algorithm is perfectly satisfactory in the cross-validation state vis-à-vis the testing condition (38). The reason behind this fact is that in the cross-validation procedure, bagged tree crosses the trained complex DTs, integrates their overlapped branches and nodes and accumulatively analyses their performance (68). Such a collective approach enhances competitively the bagged tree performance over the testing state. This is a condition in which the performance of the trained complex DTs is computed in the standalone manner and arithmetically averaged. The pattern of the dots also reflects an interesting trend in the training and learning of the bagged tree performance. The classification error has dropped sharply from the first DT training to the middle, reaching to the minimum at the 90th tree and then trivially increased with a monotonous regularity to the end. As a result, it can be inferred that 90 complex trees are sufficient and learning more leads to an overtraining problem.

Developing four types of DTs, the accuracy of the developed classification algorithms is assessed via running predictions on the testing dataset. Generally, accuracy is the most significant element in comparing various algorithms since this feature represents the functionality of the learning schema in the identification of the correct classes (68). It is evaluated by contrasting the predicted classes of the test data with their counterparts in the actual data. If the accuracy is found reliable,

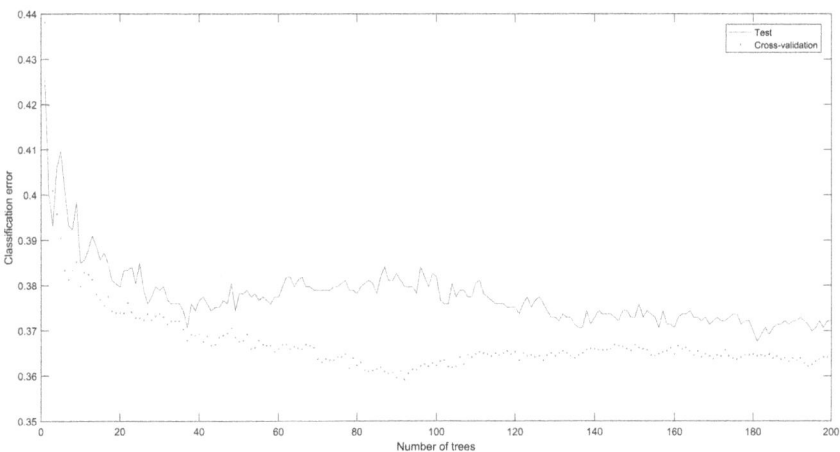

Figure 5.14 Classification Errors of the Trained Bagged Tree

the DT could be utilised for classification, prediction or objective function development for optimisation purposes. Otherwise, the problem should be investigated and diagonal activities should be adopted (38).

Therefore, the confusion matrix plots were, first, depicted to understand how the developed classifiers performed in each class (Figure 5.15). The confusion matrix assists in identifying the areas where the classifier has performed strongly or poorly. For each confusion matrix, the rows show the true (actual) classes and the columns show the predicted classes (70). The diagonal cells show where the true classes and predicted classes match. If these cells display high percentages (50% and more), the classifier has performed well and classified the observations correctly. On the contrary, the cells indicating less than 50% point to the misclassified match between the true class and the predicted class (71).

The confusion matrix for simple tree shows a good performance for 'low' and a relative performance for 'excessive' classes, whereas it performs very poorly for classifying high energy consumption class and zero performance for 'medium'

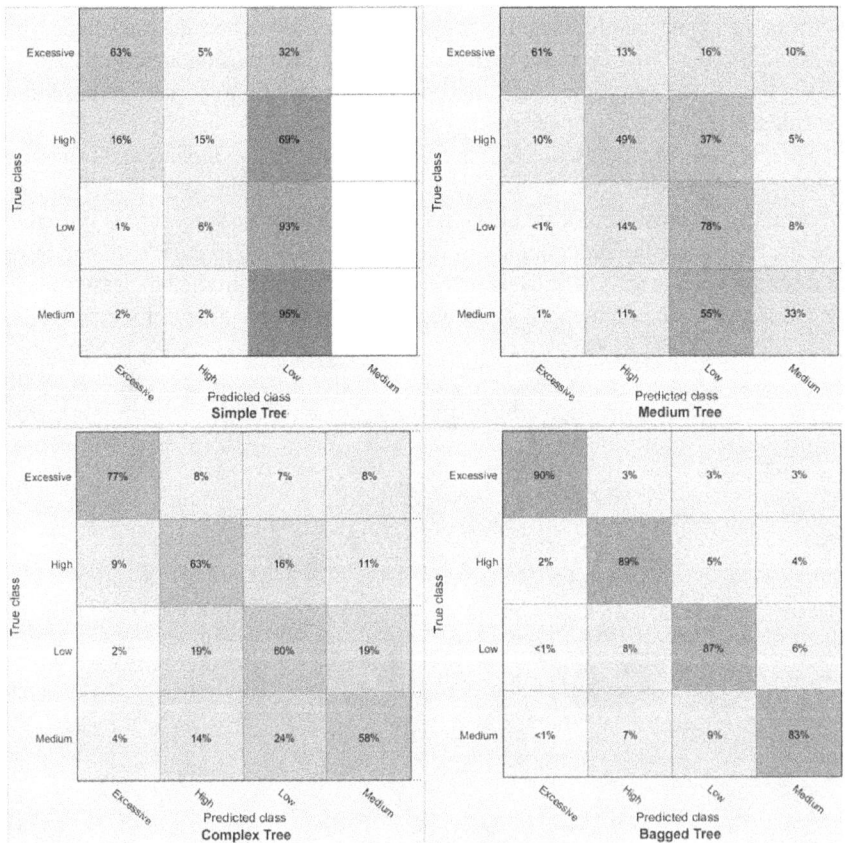

Figure 5.15 The Confusion Matrix for Four Developed DTs

class. This observation is understandable since the developed tree was very limited and could not traverse to the medium leaf nodes (see Figure 5.11). Medium and complex trees indicate a gradual growth in the classification performance, yet, are far from the acceptable level. According to Yu, Haghighat (60), accuracy of higher than 80% indicates an acceptable level of reliability for a classification algorithm. Bagged tree, in this research, classified the energy dataset with accuracy of 83%, 87%, 89% and 90% for the classes of medium, low, high and excessive.

Table 5.8 presents an average of classification performance for all DTs along with the prediction speed and training time required for learning each algorithm. It is clearly deduced that the bagged tree outperforms significantly the other classifiers with an average accuracy of 87.30%. However, with respect to the time and speed criteria, this algorithm holds the last rank. This result seems to be reasonable due to the highly complicated and integrated structure of the bagged tree as compared to the competitors (71). If the functionality of this algorithm is concerned, looking at Table 5.9, this classifier can address the energy-wise decision-making for the number of observations ranging from 800 to more than 1000 in seconds. It should be also acknowledged that there are some misclassified observations (Table 5.9), but these are, altogether, less than 15% of the observations, which is acceptable based on similar studies in the literature (3, 38, 60). Therefore, the developed bagged tree, composed from an integrated bag of 90 complex trees, was proven to be the first classifier for the energy dataset.

Table 5.8 Performance Summary for the Classification Algorithms

No.	Decision Tree Algorithm Types	Prediction Speed (OBS/Sec)	Training Time (Sec)	Accuracy
1	Simple Tree	~810000	0.40035	43%
2	Medium Tree	~1000000	0.21317	55.00%
3	Complex Tree	~930000	0.22943	64.50%
4	**Bagged Tree**	**~36000**	**2.4256**	**87.30%**

Table 5.9 Number of Observations and Data Ranges for Each Class

No. of Observations for Each Class		Predicted				Data Ranges (kWh)
		Low	Medium	High	Excessive	
Actual	Low	**959**	89	65	37	0–888
	Medium	61	**841**	45	38	888–2208
	High	84	74	**1073**	37	2208–4244
	Excessive	4	5	26	**997**	4244–Inf.

5.4.4   *Hybrid Objective Function Development*

As mentioned in Sections 5.4.1 and 5.4.3.1, single model objective functions are widely applied in the structures of prediction and classification algorithms for energy optimisations. Nevertheless, developing a unified objective function that is capable of handling both continuous and discrete datasets perfectly could alleviate the issue of qualitative and quantitative variable inclusion in a dataset. Therefore, the procedure of hybrid objective function development was started with running ANN and DT on the continuous and discrete parameters. The configurations of these algorithms were obtained from Sections 5.4.2 and 5.4.3. The procedure was then followed by weighted averaging the output of these two separately trained models in a hybrid function (72).

$$\bar{f}(\vec{x}) = \sum_k w_k f_k \vec{x},$$
(5.6)

where k is the index for each algorithm and w denotes the weight of $f(x)$ for each objective function.

C4.5 algorithm, according to the concept of entropy, was used to construct the bagged tree, including the seven discrete variables of wall, insulation roof and floor materials, glazing type and the types of main space heating and cooling systems. Progressively, the generic configuration of the algorithm is similar to the bagged DT developed for the whole dataset in the previous section. However, based on the weighted average output of the hybrid function (Equation 5.6), the final result of each function should be the same. On one hand, results falling to the classified level of energy consumption of buildings such as low, medium, high and excessive could not be arithmetically averaged with the results of ANN model, which are numerical and continuous. On the other hand, using transformation techniques may lead to randomness or losing integrity (58). Furthermore, with respect to the energy optimisation purposes, since optimisation algorithms generally work better with numerical outputs (73), it is of high priority to obtain the actual numerical level of energy consumption from the hybrid objective function.

Hence, taking advantage of the flexibility of C4.5 in handling numerical outputs, it was intended to apply standard deviation reduction instead of information-gain to calculate the homogeneity of the splits of DT (74). In this method, standard deviation is zero if the splits of DTs are completely homogenous. In fact, the decrease in standard deviation, after a dataset is split on an attribute, shapes the process of DT construction.

$$SP(X) = \sum_{D \in X} A(D) S(D),$$
(5.7)

where $SP(X)$ indicates the split of data based on the attribute (A) of standard deviation of D, based on the below.

$$S(D) = \sqrt{\frac{\Sigma(x-\mu)^2}{n}}$$
(5.8)

Likewise, the attribute with the largest standard deviation reduction is selected as the split attribute for each tree node. Considering the optimum configuration of DTs obtained from the classification section, the algorithm was set to train 100 complex trees and the test errors and cross-validation errors were computed (Figure 5.16).

It can be seen in Figure 5.16 that the bagged tree performs well in the cross-validation state as compared to the testing condition. The bottom trend shows the lowest error around 1.6 from the tenth complex tree onward, without a significant fluctuation toward the end of the training trees. However, this performance is noticeably weaker than that of bagged tree conducted for the whole dataset (see Figure 5.16) where the lowest error was 0.36. This observation may be in light of the decrease in the number of parameters from 13 to 7.[5] Overall, AI algorithms deliver greater results when they benefit from more variables in prediction and classification (75). Henceforth, this gap should be filled with advanced techniques of machine learning to enhance accuracy while reducing the number of parameters.

In this respect, applying the standard deviation in the structure of the bagged tree opens a new window of opportunity in using the ensemble regularisation technique. It is a process of removing weak learners from the DT structure and improving the performance in a way such that fewer trees are required to train the algorithm (68). This feature is also a significant achievement in the hybrid function as the time of the training could be considerably reduced, raising the speed of the optimisation engine. The regularisation procedure specifies well-trained learner weights that could minimise the errors in Equation (5.9):

$$Reg\ DT = \sum_{n=1}^{N} W_n g\left(\left(\sum_{t=1}^{T} \alpha_t h_t\left(x_n\right)\right), y_n\right) + \lambda \sum_{t=1}^{T} |\alpha_t|, \qquad (5.9)$$

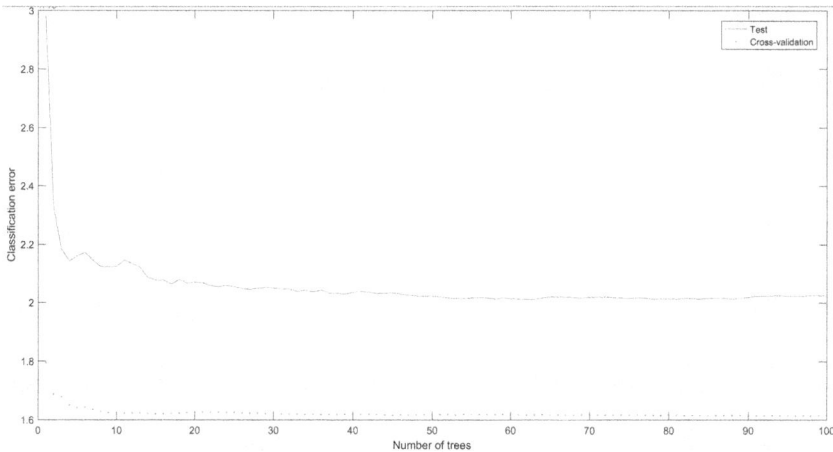

Figure 5.16 Performance Error of the Trained Bagged Tree in Hybrid Model

where α_t is the optimal set of learner weights, $\lambda \geq 0$ is the lasso parameter, h_t is a weak learner in the ensemble trained on N observations with predictors of x_n, responses of y_n, and weights of w_n and $g(f,y) = (f - y)^2$ is the squared error. The following matrix indicates a trade-off between λ and the trained weights (w_n).

$$0 \quad 0.836 \quad 0.198 \quad 0.47 \quad 0.111 \quad 0.264 \quad 0.627 \quad 0.148 \quad 0.352 \quad 0.836 \quad \begin{matrix} 0.515 \\ 0.344 \\ 0.266 \\ 0.158 \\ 0.001 \end{matrix}$$

According to this matrix, the fifth element (0.111) was found to be the smallest in the flat range, and the ensemble was reduced using the shrink method through the weighted average of the fifth element. The minimised α_t could be achieved with minimising MSE of the above equation. Usually, an optimal range could be found in which the accuracy of the regularised ensemble is better or comparable to that of the full ensemble without regularisation. In this process, if a learner's weight α_t is calculated to be 0, this learner is excluded from the regularised ensemble. In the end, an ensemble with improved accuracy and fewer learners (in comparison with unregularised ensemble) is obtained. As a result, the reduced bagged tree contained 16 complex trees in its structure along with generating approximately 0.8 cross-validated MSE (Figure 5.17). This reduced ensemble gives low loss while using far fewer trees.

Similar to the DT model development for qualitative data (categorical parameters), the ANN model was also developed upon quantitative data (continuous parameters), including ceiling height, building orientation, lighting, window to

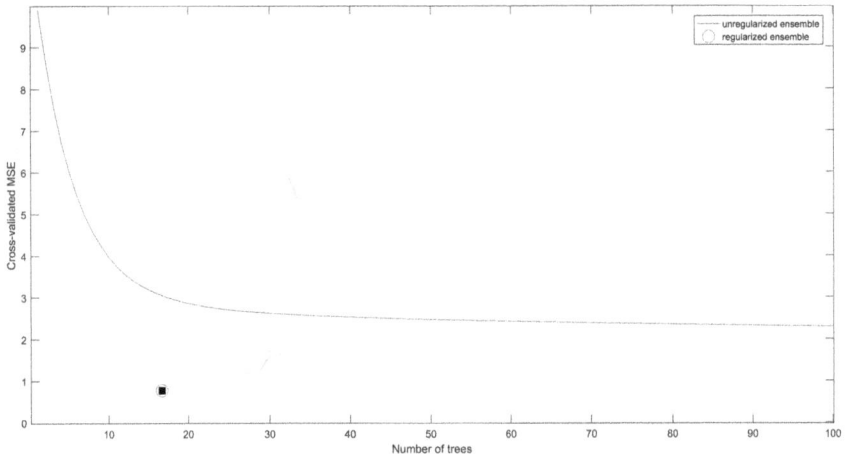

Figure 5.17 Regularised vs. Unregularised Ensemble in the Hybrid Model

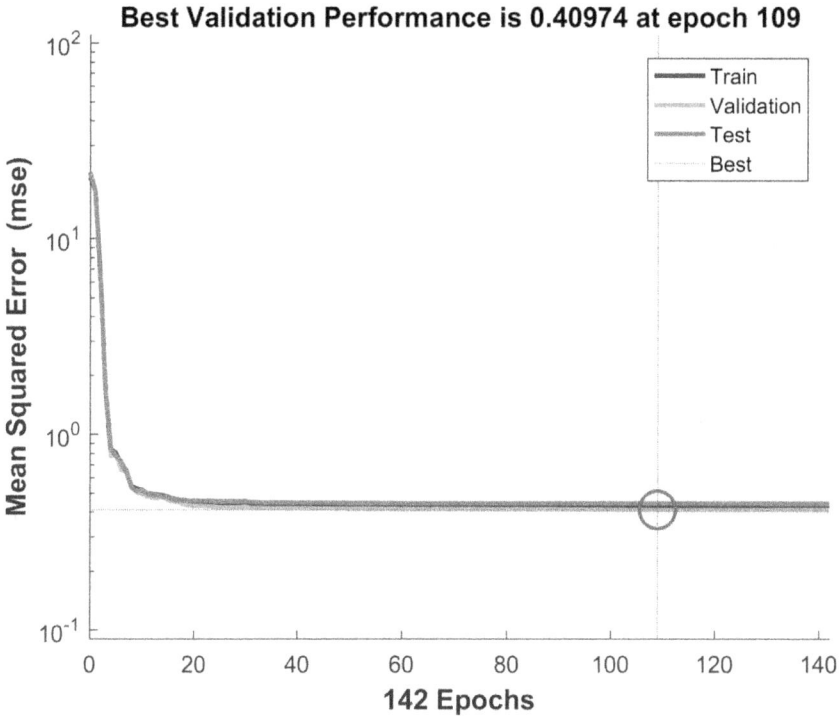

Figure 5.18 Validation Performance of ANN in the Hybrid Model

wall ratio and metres squared of rooms heated and cooled. It was made based on the configurations of ANN, identified from Section 5.4.2. Therefore, this ANN was structured with 6, 7 and 1 neurons in the input, hidden and output layers by comprising 70%, 15% and 15% of data for training, testing and validating, respectively. Hyperbolic tangent function was the activation function chosen for the input layer. Sigmoid transfer function was applied between the hidden layer and the output layer and the Levenberg-Marquardt back propagation algorithm was set as the learning algorithm. The model was set again to 1000 iterations and the best validation performance was recorded at the 109th iteration with MSE of 0.40974 (Figure 5.18).

This ANN model was trained up to 142 epochs and stopped on the rule of six consecutive runs without any decrease in the performance error. In addition, Figure 5.19 depicts the regression performance of this ANN for the different batches of training, validation, test and the overall states. These R models with figures higher than 0.75 also justify the significant correlation of the predicted outputs vs. the actual values. It is worth mentioning that the training models of DA and ANN are followed based on their own procedures. However, this is a kind of limitation in

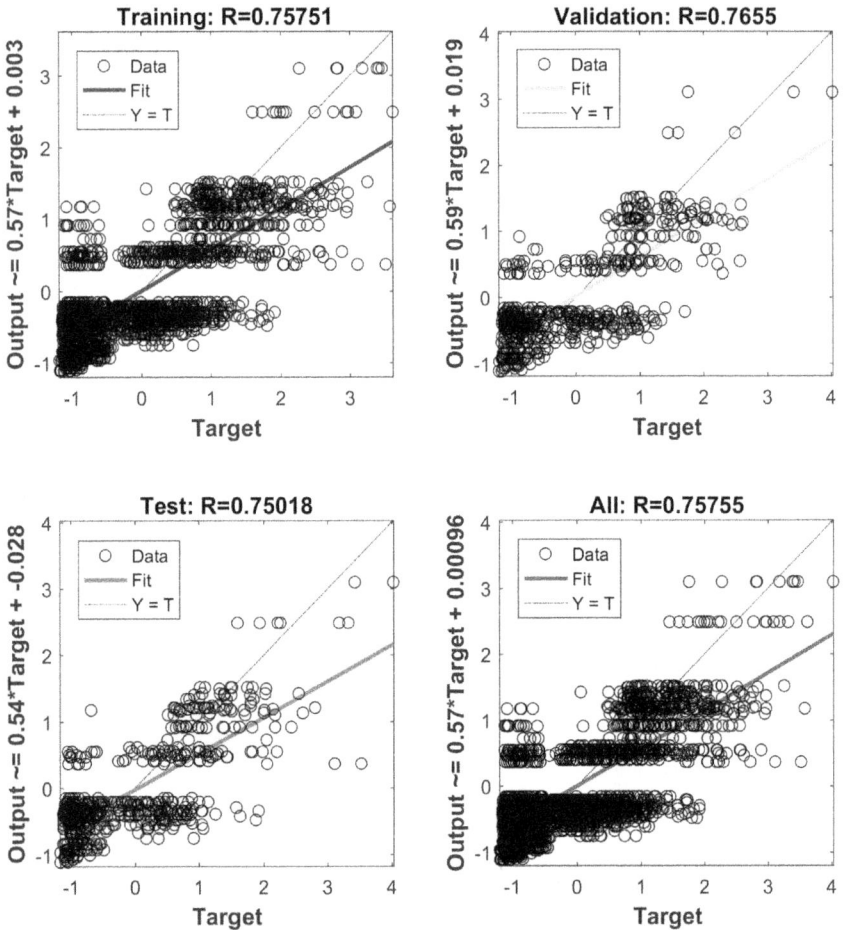

Figure 5.19 Regression Test of Hybrid ANN

developing the hybrid model. Because the simultaneous training of all data may influence each other, and this influence may even be significant.

Ultimately, referring to Equation (5.6), the hybrid model was composed from the ensemble of the bagged tree and ANN models covering both continuous and discrete parameters in one objective function (Figure 5.20). Since algorithm generation, MSE was considered the main accuracy driver; this criterion was also employed in demonstrating the improved accuracy of the hybrid model. The MSE of the hybrid model was computed approximately at 0.6, which is very low in the error rates (76). In order to validate the improved performance of the hybrid model against single objective models, first, the normalised values of the predictions of single ANN were figured out upon the whole dataset. Second, the performance

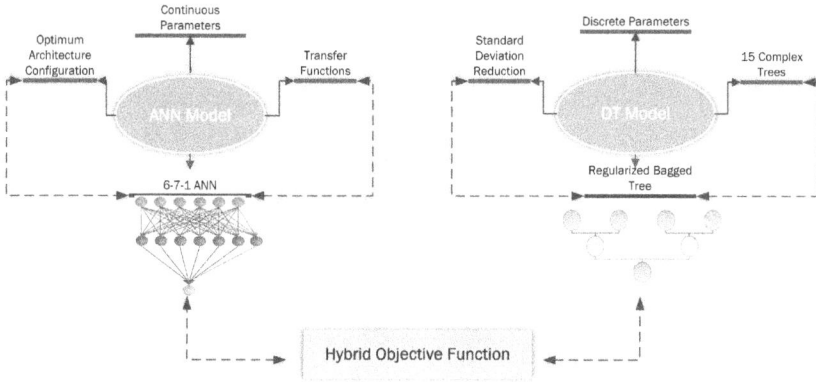

Figure 5.20 Conceptual Structure of the Hybrid Model

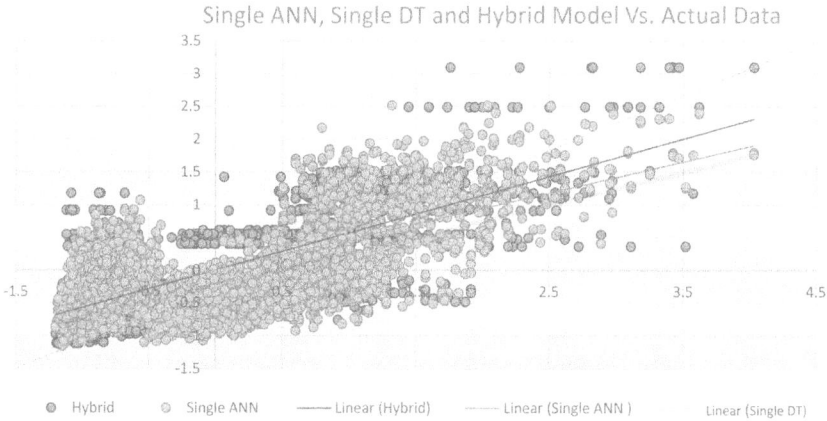

Figure 5.21 Normalised Predictive Performance of Single ANN, DT and Hybrid Model vs. Normalised Actual Energy Data

error of a single DT on the whole dataset was plotted. Finally, the associated results along with the performance of hybrid models were illustrated vis-à-vis the normalised actual outputs of energy consumption indicating the comparative performance of each model. As shown in Figure 5.21, the approximate linear trend line of the normalised values predicted by the hybrid model matches better with the equality state in comparison to that of single ANN and DT models. This observation confirms the superior performance of the hybrid model in generating predictive data as close as possible to the baseline data and provides more robust objective function.

5.5   Summary

This chapter provided efforts in fulfilling one of the very crucial objectives of this research book by developing AI algorithms of building energy prediction and classification. Driven by the literature and Delphi study in Chapter 4 with regard to identifying the significant parameters of building energy consumption, a comprehensive dataset including discrete and continuous parameters was generated via techniques of parametric simulation and further normalised through metaheuristic data size reduction. Prediction and classification algorithms ANN and DT were run separately and integrally on the dataset and a homogenous objective function model was successfully developed and tested. As a result, a hybrid algorithm of prediction and classification was established for generating energy consumption prediction with very low error and high accuracy. This paves the way for presenting a powerful engine for building energy optimisation. The outcome is an integrated objective function containing both qualitative and quantitative variables of building energy consumption, without affecting data consistency or requiring any data transformation procedures.

Notes

1 Full details can be found in Section 5.2.2.
2 Thirteen factorial simulations = $(13*12*11*10*9*8*7*6*5*4*3*2*1)$ = 6,227,020,800 in the case that each parameter has only one attribute!
3 Number of observations refers to each batch of inputs with 13 variables.
4 Datasets totalling 4435, including 13 inputs (the variables) leading to the output (annual energy consumption) consisting of 1053, 1138, 1114 and 1130 data for the cities of Sydney, Phoenix, Kuala Lumpur and Moscow, respectively.
5 Having seven categorical parameters in DT and six continuous parameters in ANN.

References

1 Karlsson J, Roos A, Karlsson B. Building and climate influence on the balance temperature of buildings. Building and Environment. 2003;38(1):75–81.
2 Nguyen A-T, Reiter S, Rigo P. A review on simulation-based optimization methods applied to building performance analysis. Applied Energy. 2014;113:1043–58.
3 Dounis AI. Artificial intelligence for energy conservation in buildings. Advances in Building Energy Research. 2010;4(1):267–99.
4 Shaikh PH, Nor NBM, Nallagownden P, Elamvazuthi I, Ibrahim T. A review on optimized control systems for building energy and comfort management of smart sustainable buildings. Renewable and Sustainable Energy Reviews. 2014;34:409–29.
5 ASHRAE. Standard 90.1–2007, energy standard for buildings except low-rise residential buildings, SI edition. American Society of Heating, Refrigerating and Air-Conditioning Engineers, Atlanta, GA; 2007.
6 Kottek M, Grieser J, Beck C, Rudolf B, Rubel F. World map of the Köppen-Geiger climate classification updated. Meteorologische Zeitschrift. 2006;15(3):259–63.
7 Michalek J, Choudhary R, Papalambros P. Architectural layout design optimization. Engineering Optimization. 2002;34(5):461–84.
8 Yu Z, Fung BCM, Haghighat F, Yoshino H, Morofsky E. A systematic procedure to study the influence of occupant behavior on building energy consumption. Energy and Buildings. 2011;43(6):1409–17.

9   Banihashemi S, Tabadkani A, Hosseini MR. Modular coordination-based generative algorithm to optimize construction waste. Procedia Engineering. 2017;180:631–9.

10 Shakouri M, Banihashemi S. Developing an empirical predictive energy-rating model for windows by using Artificial Neural Network. International Journal of Green Energy. 2012. DOI: 10.1080/15435075.2012.738451.

11 CIBSE GA. Environmental design. The Chartered Institution of Building Services Engineers; 2006.

12 Gros A, Bozonnet E, Inard C, Musy M. Simulation tools to assess microclimate and building energy: A case study on the design of a new district. Energy and Buildings. 2016;114:112–22.

13 DOE. Weather data source: U.S. Department of Energy; 2014. Available from: http://doe2.com/index_wth.html.

14 Ebrahimpour A, Maerefat M. A method for generation of typical meteorological year. Energy Conversion and Management. 2010;51(3):410–7.

15 World meteorological organization [Internet]; 2016. Available from: www.wmo.int/datastat/wmodata_en.html.

16 Shi X, Tian Z, Chen W, Si B, Jin X. A review on building energy efficient design optimization from the perspective of architects. Renewable and Sustainable Energy Reviews. 2016;65:872–84.

17 Seongchan K, Jeong-Han W, editors. Analysis of the differences in energy simulation results between Building Information Modeling (BIM)-based simulation method and the detailed simulation method. Simulation Conference (WSC), Proceedings of the 2011 Winter; December 11–14, 2011.

18 Jaffal I, Inard C, Ghiaus C. Fast method to predict building heating demand based on the design of experiments. Energy and Buildings. 2009;41(6):669–77.

19 Pisello AL, Goretti M, Cotana F, editors. Building energy efficiency assessment by integrated strategies: Dynamic simulation, sensitivity analysis and experimental activity. Proceedings of Third International Conference on Applied Energy; 2011.

20 van Treeck C, Rank E. Dimensional reduction of 3D building models using graph theory and its application in building energy simulation. Engineering with Computers. 2007;23(2):109–22.

21 Eriksson L, Johansson E, Kettaneh-Wold N, Wikström C, Wold S. Design of experiments. Principles and Applications. 2000:172–4.

22 Bailey RA. Design of comparative experiments. Cambridge Series in Statistical and Probabilistic Mathematics (No. 35): Cambridge University Press; 2008.

23 Mara TA, Tarantola S, editors. Application of global sensitivity analysis of model output to building thermal simulations. In: Building simulation. Springer; 2008.

24 Montgomery DC. Design and analysis of experiments: John Wiley & Sons; 2008.

25 Leach N. Digital morphogenesis. Architectural Design. 2009;79(1):32–7.

26 Bollmann D, Bonfiglio A. Design constraint systems: A generative approach to architecture. International Journal of Architectural Computing. 2013;11(1):37–64.

27 Dino I. Creative design exploration by parametric generative systems in architecture. METU Journal of Faculty of Architecture. 2012;29(1):207–24.

28 Liu Y-C, Chakrabarti A, Bligh T. Towards an 'ideal' approach for concept generation. Design Studies. 2003;24(4):341–55.

29 Abotaleb I, Nassar K, Hosny O. Layout optimization of construction site facilities with dynamic freeform geometric representations. Automation in Construction. 2016;66:15–28.

30 Talbi E-G. Metaheuristics: From design to implementation: John Wiley & Sons; 2009.

31 Gendreau M, Potvin J-Y. Handbook of metaheuristics: Springer; 2010.

32 Weiss NA, Weiss CA. Introductory statistics: Pearson Education London; 2012.

33 Neustadt LW. Optimization: A theory of necessary conditions: Princeton University Press; 2015.

34 Machairas V, Tsangrassoulis A, Axarli K. Algorithms for optimization of building design: A review. Renewable and Sustainable Energy Reviews. 2014;31:101–12.

35 Blum C, Roli A. Metaheuristics in combinatorial optimization: Overview and conceptual comparison. ACM Computing Surveys (CSUR). 2003;35(3):268–308.

36 Demuth HB, Beale MH, De Jess O, Hagan MT. Neural network design: Martin Hagan; 2014.

37 Yuce B, Rezgui Y, Mourshed M. ANN-GA smart appliance scheduling for optimised energy management in the domestic sector. Energy and Buildings. 2016;111:311–25.

38 Iqbal M, Azam M, Naeem M, Khwaja A, Anpalagan A. Optimization classification, algorithms and tools for renewable energy: A review. Renewable and Sustainable Energy Reviews. 2014;39:640–54.

39 Ahmed A, Korres NE, Ploennigs J, Elhadi H, Menzel K. Mining building performance data for energy-efficient operation. Advanced Engineering Informatics. 2011;25(2):341–54.

40 Caruana R, Niculescu-Mizil A, editors. An empirical comparison of supervised learning algorithms. Proceedings of the 23rd International Conference on Machine Learning, ACM; 2006.

41 Flores JA. Focus on artificial neural networks. Nova Science; 2011.

42 Kruse R, Borgelt C, Klawonn F, Moewes C, Steinbrecher M, Held P. Multi-layer perceptrons. In: Computational intelligence. Springer; 2013. p. 47–81.

43 Shahidehpour M, Yamin H, Li Z. Market operation in electric power systems. Wiley-IEEE Press; 2002.

44 Ramedani Z, Omid M, Keyhani A. Modeling solar energy potential in Tehran Province using artificial neural networks. International Journal of Green Energy. 2012;10(4):427–41.

45 Alsmadi MKS, Omar KB, Noah SA. Back propagation algorithm: The best algorithm among the multi-layer perceptron algorithm. IJCSNS International Journal of Computer Science and Network Security. 2009;9(4):378–83.

46 Zhang J-R, Zhang J, Lok T-M, Lyu MR. A hybrid particle swarm optimization: Back-propagation algorithm for feedforward neural network training. Applied Mathematics and Computation. 2007;185(2):1026–37.

47 MacKay DJ. Bayesian interpolation. Neural Computation. 1992;4(3):415–47.

48 Møller MF. A scaled conjugate gradient algorithm for fast supervised learning. Neural Networks. 1993;6(4):525–33.

49 Levenberg K. A method for the solution of certain non-linear problems in least squares. Quarterly of Applied Mathematics. 1944;2(2):164–8.

50 Battiti R. First-and second-order methods for learning: Between steepest descent and Newton's method. Neural Computation. 1992;4(2):141–66.

51 Riedmiller M, Braun H, editors. A direct adaptive method for faster backpropagation learning: The RPROP algorithm. Neural Networks, 1993, IEEE International Conference On, IEEE; 1993.

52 Baldi P. Gradient descent learning algorithm overview: A general dynamical systems perspective. IEEE Transactions on Neural Networks. 1995;6(1):182–95.

53 Foucquier A, Robert S, Suard F, Stéphan L, Jay A. State of the art in building modelling and energy performances prediction: A review. Renewable and Sustainable Energy Reviews. 2013;23:272–88.

54 Evins R. A review of computational optimisation methods applied to sustainable building design. Renewable and Sustainable Energy Reviews. 2013;22:230–45.

55 Ahmad A, Hassan M, Abdullah M, Rahman H, Hussin F, Abdullah H, et al. A review on applications of ANN and SVM for building electrical energy consumption forecasting. Renewable and Sustainable Energy Reviews. 2014;33:102–9.

56 Rawat R, Patel JK, Manry MT, editors. Minimizing validation error with respect to network size and number of training epochs. Neural Networks (IJCNN), The 2013 International Joint Conference on, IEEE; 2013.

57 Cohen PR, Feigenbaum EA. The handbook of artificial intelligence: Butterworth-Heinemann; 2014.

58 Lebart L, editor. Correspondence analysis. Data Science, Classification, and Related Methods: Proceedings of the Fifth Conference of the International Federation of Classification Societies (IFCS-96), Kobe, Japan, March 27–30, 1996, Springer Science & Business Media; 2013.

59 Kotsiantis SB, Zaharakis I, Pintelas P. Supervised machine learning: A review of classification techniques. IOS Press; 2007.

60 Yu Z, Haghighat F, Fung BC, Yoshino H. A decision tree method for building energy demand modeling. Energy and Buildings. 2010;42(10):1637–46.

61 Kotsiantis SB. Decision trees: A recent overview. Artificial Intelligence Review. 2013;39(4):261–83.

62 Kohavi R, Quinlan JR, editors. Data mining tasks and methods: Classification: Decision-tree discovery. In: Handbook of data mining and knowledge discovery. Oxford University Press, Inc; 2002.

63 Muniyandi AP, Rajeswari R, Rajaram R. Network anomaly detection by cascading k-Means clustering and C4. 5 decision tree algorithm. Procedia Engineering. 2012;30:174–82.

64 Quinlan JR. Induction of decision trees. Machine Learning. 1986;1(1):81–106.

65 Quinlan JR. C4. 5: Programs for machine learning: Elsevier; 2014.

66 Shannon CE, Weaver W. The mathematical theory of communication: University of Illinois Press; 2015.

67 Han J, Pei J, Kamber M. Data mining: Concepts and techniques: Elsevier; 2011.

68 Tang J, Alelyani S, Liu H. Feature selection for classification: A review. In: Aggarwal C, editor. Data classification: Algorithms and applications. CRC Press In Chapman & Hall/CRC Data Mining and Knowledge Discovery Series; 2014.

69 Breiman L. Random forests. Machine Learning. 2001;45(1):5–32.

70 Roiger RJ. Data mining: A tutorial-based primer: CRC Press; 2017.

71 Rokach L, Maimon O. Data mining with decision trees: Theory and applications: World Scientific; 2014.

72 Merkwirth C, Wichard J, Ogorzałek M. ENTOOL-A Matlab toolbox for regression, classification and active learning. Jagiellonian University; 2007.

73 Delgarm N, Sajadi B, Kowsary F, Delgarm S. Multi-objective optimization of the building energy performance: A simulation-based approach by means of particle swarm optimization (PSO). Applied Energy. 2016;170:293–303.

74 Tso GK, Yau KK. Predicting electricity energy consumption: A comparison of regression analysis, decision tree and neural networks. Energy. 2007;32(9):1761–8.

75 Gandomi AH, Yang X-S, Talatahari S, Alavi AH. Metaheuristic applications in structures and infrastructures: Newnes; 2013.

76 Tao F, Zhang L, Laili Y. Configurable intelligent optimization algorithm. Springer; 2014.

6 BIM-Inherited EED Framework Development and Verification

6.1 Optimisation Procedure

Optimisation procedure in this book commenced with identifying the 13 parameters as the optimisation variables (see Table 5.1) and the hybrid objective as the fitness function for optimisation. As discussed in previous chapters, GA (genetic algorithm) was selected for optimisation in this research. This is in light of its superior flexibility in handling the hybrid objective function, reliability and precision in finding the optimum values, fast reaction for big datasets and its popularity among the scientific community (1). GA operator initiates the optimisation procedure by corresponding the variables and their values with chromosomes in Matlab. GA changes the values of parameters into the binary codes and then randomly generates the initial population (2). The distances of chromosomes in the initial population from each other are calculated and checked through the minimisation criterion. This algorithm, iteratively, improves the solutions via generating more populations by learning from the previous iterations (3). The most widely applied stopping criteria in GA optimisation is to check the convergence of populations into the single solution (4).

The algorithm can also be terminated when there is not any improvement in finding the best solution over the number of generations. In this condition, and as illustrated in Figure 6.1, two important strategies of crossing over and mutation were implemented for minimising the risk of early termination without reaching the best response (5). Crossing over creates one or more children from some selected parent chromosomes and the resulting offspring are placed into the population to cover the new range of population. Mutation alters a percentage of chromosomes to explore the full solutions and prevent GA from converging too quickly prior to sampling the entire surface (6).

According to Figure 6.1, GA worked toward the satisfied convergence of the energy optimisation goal. Figure 6.2 confirms the performance of GA in reaching the perfect convergence below 80 generations. A sharp drop could be seen during the first 20 generations of GA where the GA gets better over time and then reaches a gradual plateau. The minimum of approximately −750 becomes

DOI: 10.1201/9781003207658-6

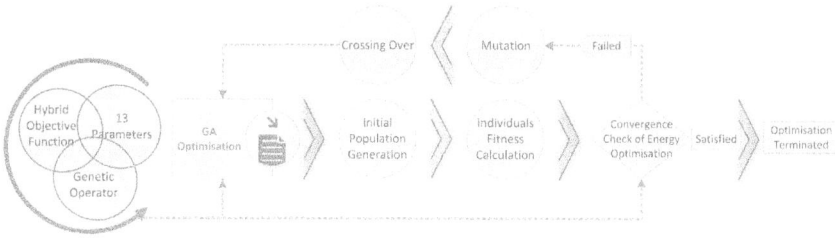

Figure 6.1 Optimisation Procedure Diagram

constant after around 20 generations, and then, strategies of mutation and crossing over are run to mock up all of the available solutions. So, as can be inferred, the height of the generation bars diminishes progressively. The trend after around 30 generations is monotonous till the complete stoppage of GA below 80 generations, and this indicates the perfect convergence of GA and its great performance in the fast track.

The average distance between individuals is also presented vis-à-vis the generations in Figure 6.3. This plot specifies how the populations are diversely scattered during optimisation (5). A population has high diversity if the average distance is large; otherwise, it has low diversity. Diversity is essential to GA because it enables the algorithm to search a larger region of the space (3). As depicted in Figure 6.3, there are higher levels of diversity among the first 30 generations, whereas it is diminishingly scattered toward the convergence in the end. This fact implies that including a wide range of solutions has led to a uniform result, confirming again the great performance of the optimisation algorithm.

6.2 Integration Framework

This section develops the framework of AI-enabled BIM-inherited EED: the package of AI (i.e. ANN, DT and GA) including prediction, classification and optimisation algorithms inbuilt in BIM. This presents a promising solution to the data integration problems and contextualises the BIM-inherited EED, which is effectively equipped with AI. The bottom line of the integration framework lies in a method built up on the dexterous capability in minimising the reinterpretation and conversion of data. The key here is keeping data in the native format as much as possible by attempting to adapt them from different sources of information and maintain the homogeneity of the optimised data. Therefore, the integration workflow of this framework was structured on five major stages of database development, database exchange, database optimisation, database

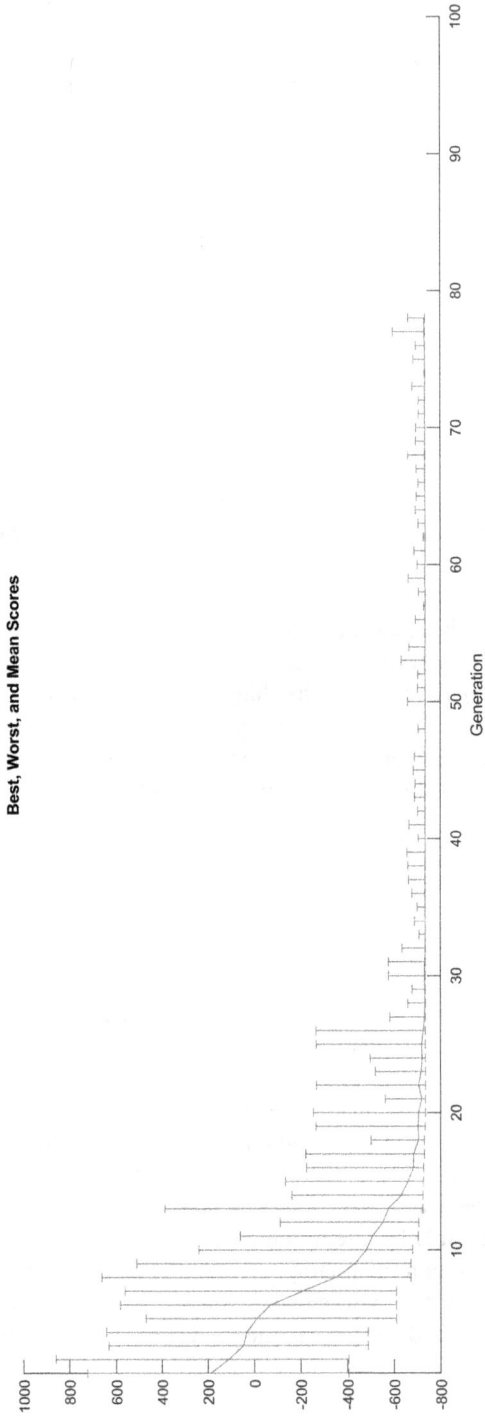

Figure 6.2 Convergence Performance of GA

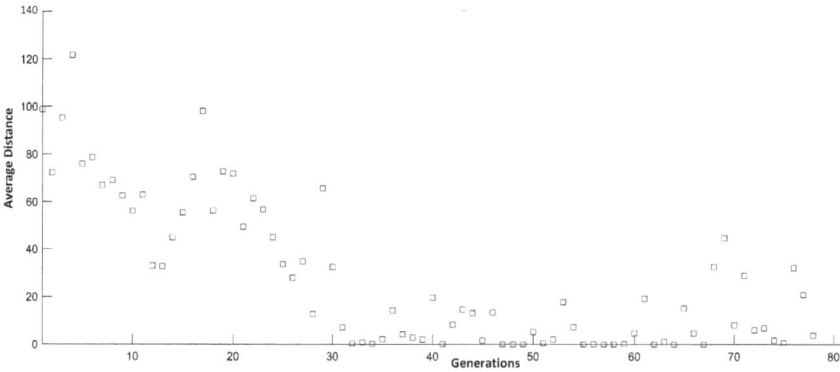

Figure 6.3 Average Distance between Individual Results

switchback and database updating, as illustrated in Figure 6.4 and discussed in the following.

6.2.1 Database Development

The first stage, database development, denotes the modelling strategies and the standard of information that should be implemented in this framework (Figure 6.4). Modelling here means to represent the real condition of the 'model elements' in the most representative and accurate manner possible (7). As discussed in Chapter 3, model elements are the components included within the model that contain the quantity, size, shape, location and orientation to represent the particular object, system or assembly (buildings in this study). Assembling these model elements together results in the model federation. The federation process is to combine the various discipline outputs such as fabric, structure, lighting, mechanical, electrical and plumbing services and, generally, all AEC deliverables into one model (8). However, in order to see the impact of design, orientation and engineering decisions on the energy optimisation, non-graphic information such as semantic and project information should also be attached to the model elements.

BIM is especially regarded as the model-based technology linked with a database of project information. It integrates semantically rich information related to the facility, including all geometric, non-geometric and functional properties during the whole lifecycle in a collection of smart model elements (9). Therefore, beyond consisting of a high level of accurate geometric information, and forming an advanced 3D graphical representation of the project, LoD 300 (see Section 3.2.7) should be dealt with priority in the database development stage. The reason for targeting this LoD is to bring the model to the construction documentation level for paving the way of the detailed energy estimation and optimisation. It further alleviates the risk of generic variable inclusion and the disassociation between the semantic values with the topological relationship. LoD 300 positively affects the building performance analysis and optimisation quality in the database

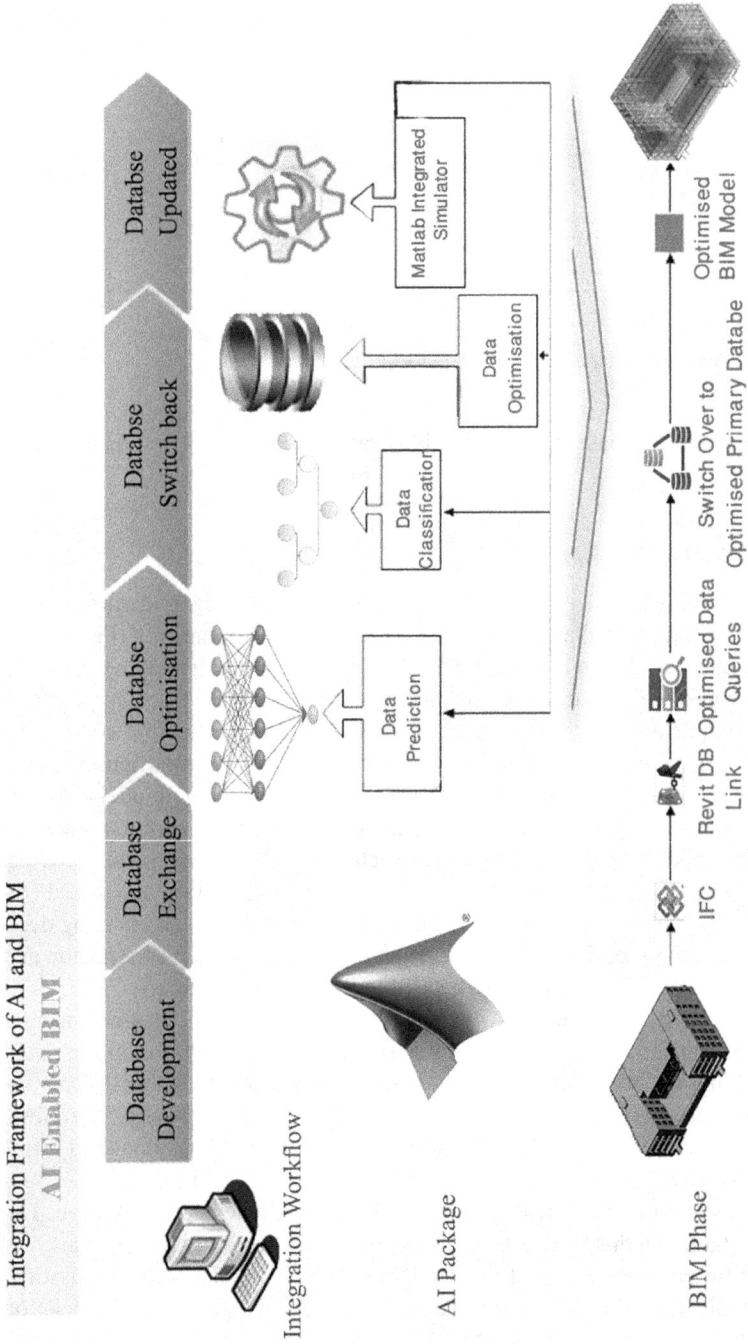

Figure 6.4 AI and BIM Integration Framework (AI-Enabled BIM-Inherited EED)

development process. It also dictates the minimum dimensional, spatial, quantitative, qualitative and other data included in a model element (10).

As the last but not least consideration in this stage, the authorised uses of LoD 300 in database development should be exerted in covering the wide range of values by BIM families. In other words, the parametric change engine of the BIM authoring tool, Revit, should be matched with the data ranges that have already been established in the optimisation algorithms. This secures the semantic association with the geometrical and topological configurations of data enrichment in BIM elements and families. This notion further ensures that once a change is made to a family or a model element in a BIM model, it is propagated throughout the database. This propagation acts according to the defined level of information in view of their compatibility with three BIM-based data enrichment criteria of topological, geometrical and semantic associations.

6.2.2   Database Exchange

Database exchange constitutes the second major stage of the framework and is run by two consecutive steps of information export with IFC format and Revit Database (DB) link. For information export, a predefined series of data drops is specified in the database exchange protocol to identify the required exchange schema (Figure 6.5). However, as a non-proprietary schema, the interoperability cannot be solely provided with IFC; rather, it requires a platform to interface with it. Through DBLink, which is a tool for Revit projects to be exported into

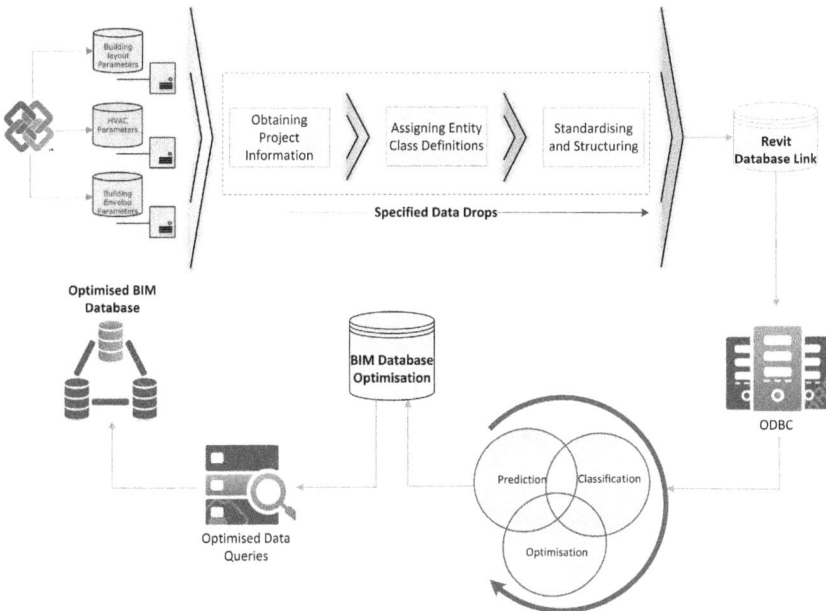

Figure 6.5 Database Exchange, Optimisation and Switching Back Process

a database, the exchanged database of the BIM model can be easily explored as the *mdb* file and can be read with the external applications. In particular, database exchange here represents the full interoperability of the BIM database and keeps its competency to communicate with multiple software products.

Practically, in this stage, first, the project information including building layout, HVAC and physical properties and building envelope parameters are extracted in the IFC format. Afterwards, property sets are assigned with entity class definitions for interrogating BIM design solutions and building knowledge ontologies. As the last step of specified data drops, the entity class definitions are standardised and structured based on IFC schema as to the enhancement of data export accuracy for a strict regime of model elements classification. Conducting all specified data drops tasks, Revit DBLink is applied to synthesise the IFC database to a readable one. The challenge in using IFC is to find a method to manipulate the database in a transparent format so that the AI package could be implemented on the database. Therefore, Revit DBLink is utilised to associate the database with Matlab via Open Database Connectivity (ODBC), which is an open standard Application Programming Interface (API) for accessing a database. By using ODBC statements in a program, the database can be accessed in a number of different formats via calling APIs for different applications such as Matlab, Access, dBase, Excel and Text (11). Hence, an industry-wide, open and neutral data format that is fast becoming the de facto standard for rich data exchange is created by an ad hoc linking database and is accessible via Matlab (see Figure 6.5).

6.2.3   *Database Optimisation*

Database optimisation conceptualises the third stage of the integration framework and is particularly designated to pinpoint the workflow of optimising the database arising from ODBC. The exchanged database presents a mixture of semantic, topological and geometric entities, which are classified based on ontology query-able format. So, it should be primarily instantiated with available values of AI package in Matlab environment, as depicted in Figure 6.6. For this reason, the open database coheres with the parameter placeholders embedded in the Matlab integrated simulator (see Figure 6.4), which, in turn, develops the verified property fields. This achievement results in a prompt reaction of the Matlab integrated simulator on calling the package of prediction, classification and optimisation algorithms. The mechanism of calling is based on the IFC property sets and entity classifications to quickly find the 13 building parameters that are the target of AI.

Highlighting the target parameters and the respected values from a long list of project information, the hybrid objective function is, first, activated. This operates on the parameters by dividing them into continuous and categorical variables. This momentum is done via the entity classifications introduced in the database exchange section. ANN and DT algorithms work on the continuous and categorical data, respectively, and the hybrid objective function aggregates and sends the results (energy estimation) to the optimisation algorithm. GA, as the optimisation engine, further processes data to the optimised level by finding the best combination of the parameters that could minimise the energy estimation value (the results coming from the hybrid objective function). Accordingly, the Matlab integrated

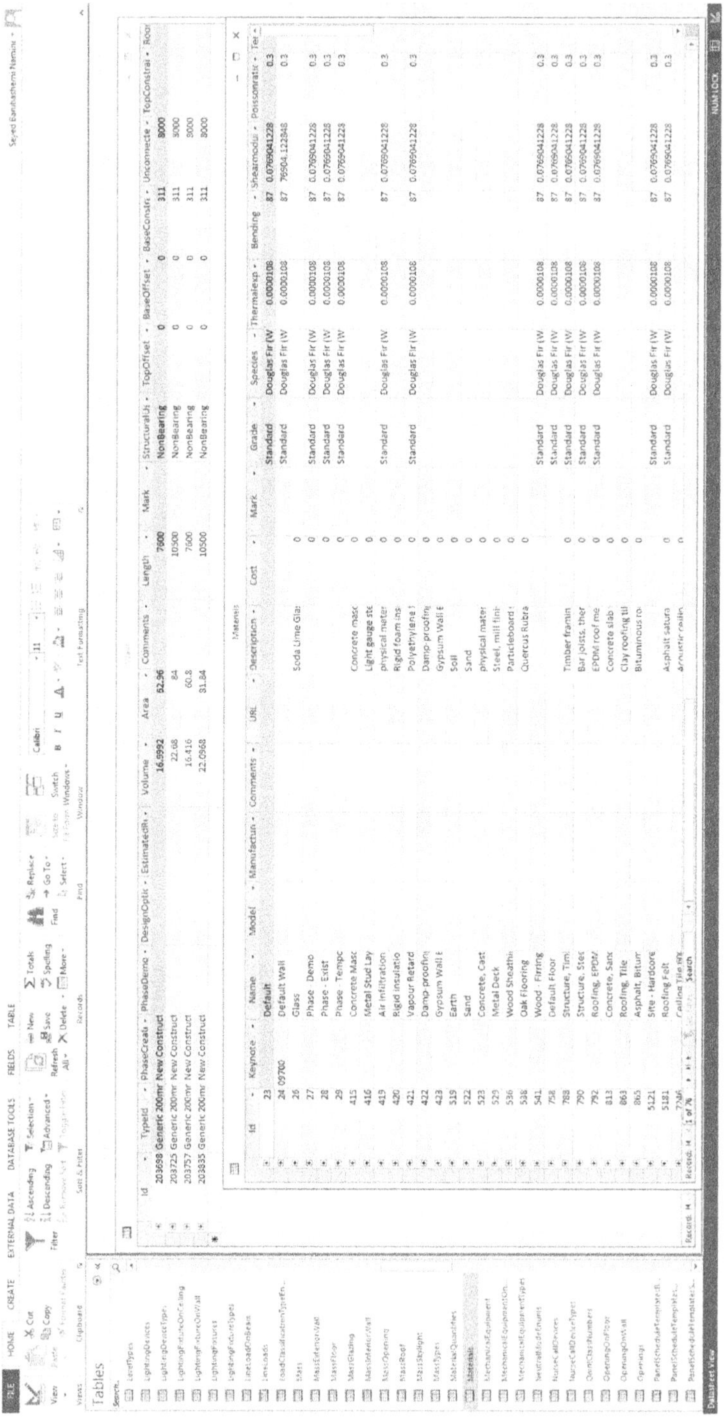

Figure 6.6 ODBC Database Structure

simulator saves the outcomes in their respected entities and creates a new database called BIM database optimisation (see Figure 6.5).

6.2.4 Database Switchback

The fourth stage of the integration framework, database switchback, ensures that this is not a one-time static export and all entities and parameters modification, outside of BIM environment, must be switched back into BIM (see Figure 6.4). It means that the changes in the values of model elements that occur as the result of the database optimisation stage should automatically correspond to their counterparts in the BIM model. Hereby, ODBC comes again to the fore by dynamically linking the optimised DB with the existing query-able entities of the database. It should also be noted that some new fields of values can be added to the database and linked to the BIM model as the shared parameters, though they are not optimisable due to being out of the realm of the target parameters.

The significant technical element of the database switchback is the optimised data query command, which provides control over the stored data in the optimised database, as depicted in Figure 6.5. This mechanism distinguishes the optimised parameters from the rest based on their defined entity classes and sets their Boolean property editor to 'on'. Query tab is then launched to glean all verified instances with the 'on' conditions and to send back the data to the BIM database. Among the original features, there is the possibility of associating the query-able data to different objects of the original IFC database, which transforms the numerical BIM model elements into a powerful navigation interface in the project construction database. This feature allows for the framework to save considerable time in browsing and understanding complex database switchback structure. This stage incorporates a mechanism for BIM to plug in the variations of the model database to its updated version.

6.2.5 Database Updated

By running the database switchback, the primary BIM model is updated with the optimised values and, hence, the last stage of the integration framework, database updated, is completed, as indicated in Figure 6.1. The technical approach here starts with overwriting the optimised values of the target parameters, which were inferred from the previous stages, with their initial values (database development stage). With respect to the suited BIM database content management, the whole data are merged into a single entity and recorded in the Revit file format. Once the database is successfully overwritten, Revit generates a report to specify the modified values and parameters.

This stage enables the object mapping between two states (pre- and post-optimisation) of the BIM database via an object-oriented modelling concept. When object-oriented programming is considered in updating the database, it effectively facilitates natural object mapping between Revit and all external database applications such as DB and ODBC. Therefore, once the required instances are populated from the prototype, the updated values are encapsulated in model element instances. Finally, the object-oriented nature of Revit allows it to automatically update the BIM model in any design view (Figure 6.7).

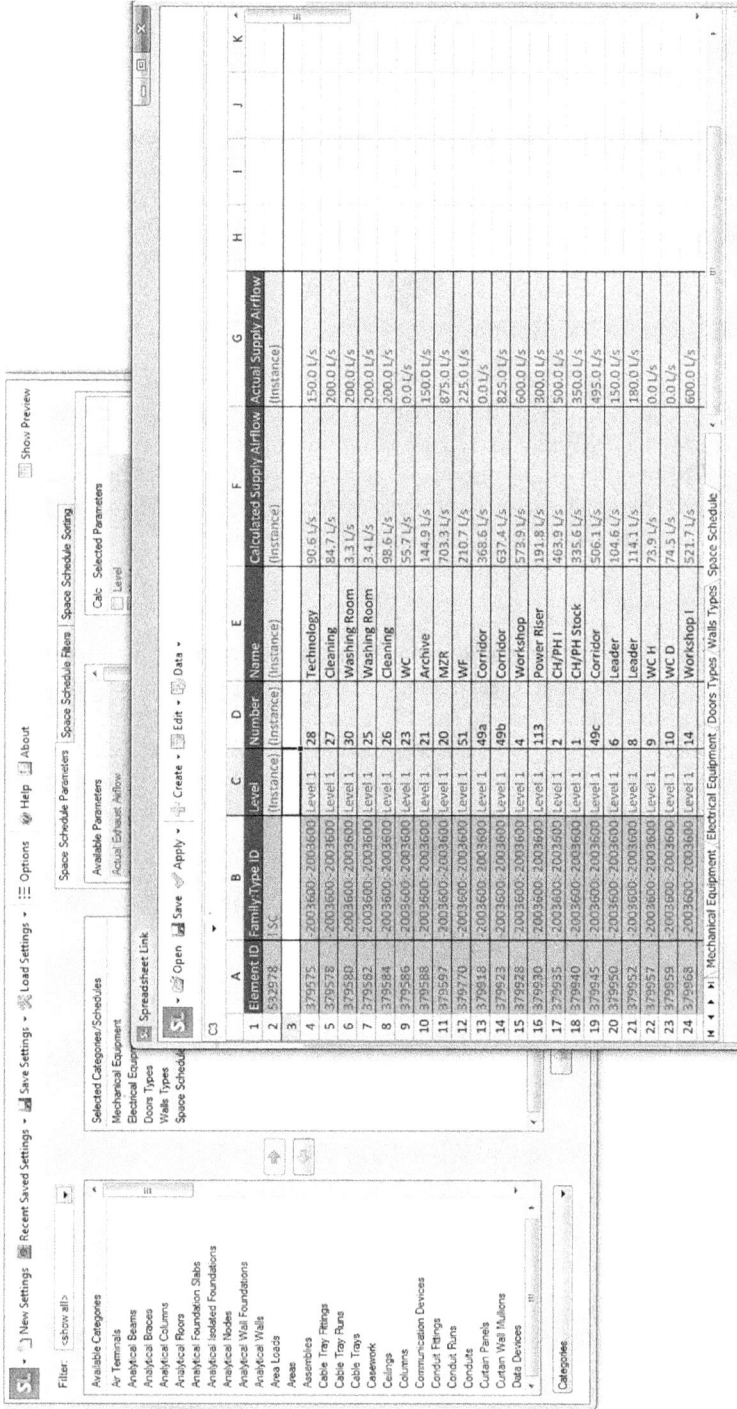

Figure 6.7 The Database Update Process

The developed framework presents an integration method for conducting BIM-inherited EED, AI-enabled BIM, which optimises the BIM model database via AI algorithms. Through this framework, the interdisciplinary data interoperability between the architectural and construction model with EED is made in CDE and the seamless environment of BIM model elements. This fact leverages the consistency in utilising building layout, physical properties, building envelop and HVAC parameters without manually simulating in external applications. Data reuse from the BIM model minimises the manual efforts in developing input data, removes the error-prone process and enables the system interface with better accuracy. However, the functionality of the framework should be tested to prove its performance.

6.3 Testing and Validation

The last part of the data analysis in this study presents a proof of concept on the reliability and workability of the developed AI algorithms and the framework of their integration (BIM-inherited EED) in optimising the energy consumption of residential buildings. For this reason and according to the adopted methodology, a case study was chosen to be analysed in its existing (baseline) and optimised states in order to indicate a statistically significant transformation.

6.3.1 Case Study

The baseline case study[1] chosen for the verification objective is a three-story building consisting of four units on each floor, which represents the conventional type of mid-rise residential apartments in Australia. Each level area is 820 m^2, summing up to a total area of 2460 m^2. The case is a real-life example of a residential building that is located on Sydney and was modelled in the BIM authoring tool: Autodesk Revit with LoD 300. Figure 6.8 shows the rendered BIM model and Figure 6.9

Figure 6.8 BIM Model of the Baseline Case Study

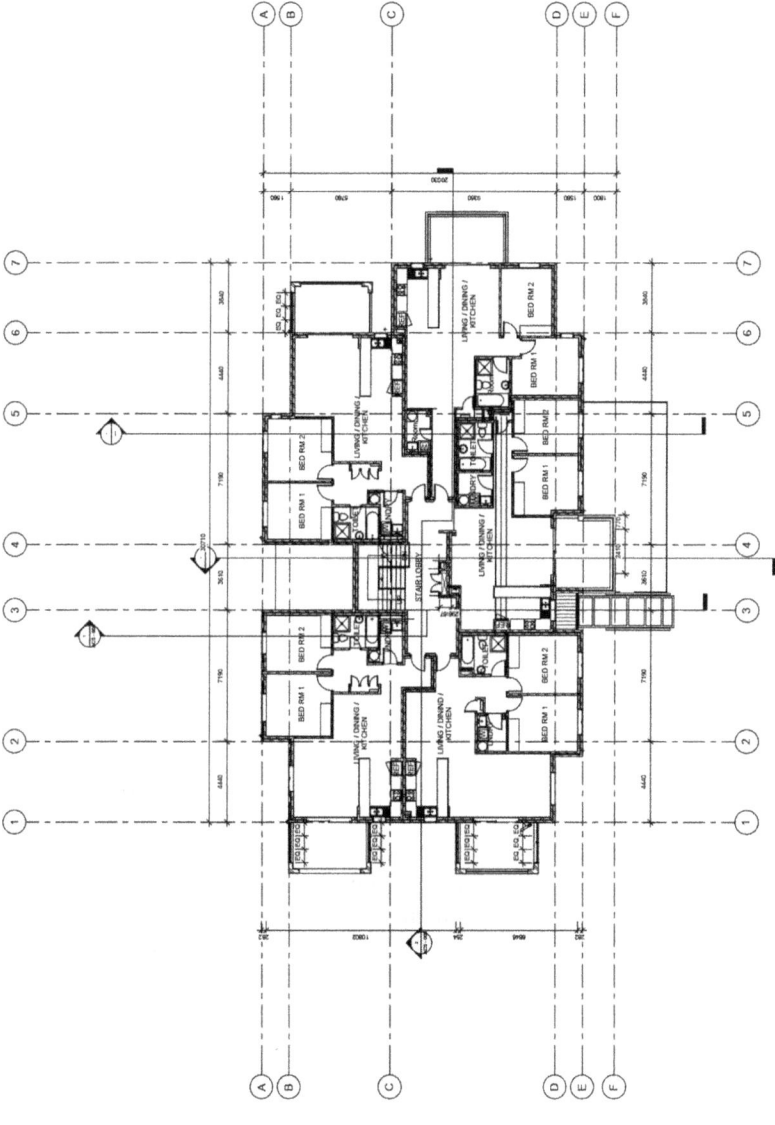

Figure 6.9 Layout of the Baseline Case Study

Table 6.1 Baseline Case Study Specifications

Construction Component	Specification	Thickness
External Wall	Double Brick Wall	270 mm
	Masonry	110 mm
	Thermal Air	50 mm
	Masonry	110 mm
	Cement Render	12 mm
Insulation	Air Terminal	50 mm
Roof	Roofing Pitched	70 mm
	Concrete Roof Tile	100 mm
	Timber Roof Truss	30 mm
Floor	Reinforced Concrete Floor	170 mm
Windows	Single Glazed	NA
Orientation	45°	NA
Ceiling Height	3000 mm	NA
Area of Rooms Cooled	725 m^2	NA
Area of Rooms Heated	725 m^2	NA
Type of Main Space Heating	Heat pump	NA
Type of Main Space Cooling	Cooling Fan	NA
Lighting	4.84 W/m^2	NA
Window to Wall Ratio	15%	NA

depicts the layout plan of the baseline case. Table 6.1 also indicates the construction of the model and its specifications.

6.3.2   Energy Simulation

For energy simulation of the baseline case in its existing state, GBS, which uses the EnergyPlus engine and is effectively inherited in Revit, was employed. Although the focus is on the design stage, some parameters such as preliminary schedules for occupancy should be taken for granted to enable the software to proceed through simulation. Hence, the model was defined as the analytical surfaces to calculate the kitchen, bedroom, bathroom and living room as the zones and their adjacencies and inter-zonal relations. For providing the thermal comfort inside the units, the thermostat was fixed between 18 and 26 °C (12) and HVAC devices were set to be activated below or above this range. Table 6.2 indicates the user profile of the zones in the building.

With respect to the specifications of the baseline case simulation (Table 6.1), double brick construction has been used for walls having a U-value of 1.5 W/m²K. Concrete roof tiles with 100 mm thickness, covered with a pitched roof with 70 mm thickness and 30 mm of timber roof truss and the total U-value of 1.61 W/m²K, has been applied for the roof of the case. Besides, a single glazed window with metal frame and a U-value of 6 W/m²K has been used as the typical materials used in residential buildings in Australia. With respect to the context of the baseline

Table 6.2 User Profile

Zone	Area (m²)	Volume (m³)	Occupancy	Activity	Comfort Band
Living Room	19	57	4	Sedentary	18–26 °C
Kitchen	10	30	4	Cooking	NA
Bathroom	4	12	1	Sedentary	NA
Bedroom	12	36	2	Sedentary	18–26 °C

case, Sydney is the most populous city of Australia and is located on −33.86° latitude and 151.2° longitude (see Table 5.3). It enjoys a temperate climate with a mild winter and has about 104 sunny days a year (13). Since Sydney never experiences extremely hot or cold days throughout a year, the loads that the designer should consider while designing for air conditioning systems are not that extreme, either (14). The climatic data of Sydney can be found in Table 5.4 as well.

6.3.3 Baseline Case Simulation Results

The simulation was run for the baseline model in Sydney and the results were recorded as depicted in Figures 6.10 and 6.11. Figure 6.10 illustrates the monthly electricity consumption and Figure 6.11 indicates the total energy consumption including electricity, hot water and gas for heat pump application. As a holistic view on these two figures, it is inferred that space cooling and heating loads have the largest share of the energy consumption drivers for the baseline model, as expected. However, the former is the most influential element for electricity consumption (Figure 6.10), and the latter plays the key role in the total energy consumption estimation (Figure 6.11). The reason is that for cooling space, cooling fans are used in this case, and these items work by electricity power. But for heating space, the heat pump system is employed, which uses another source of power: gas. This fact also causes two completely different trends for electricity and total energy consumption (i.e. electricity and gas) in which the former is U-shaped (Figure 6.10), but the latter experiences are relatively n-shaped (Figure 6.11).

Cooling load is the rate at which electricity is consumed at the cooling coil that serves one or more conditioned spaces in any cooling system, such as A/C and fans. The total building cooling load consists of sensible and latent loads. The former includes the heat transferred through the building envelope, such as walls, roof, floor, windows, doors etc., and the latter encompasses the heat generated by occupants, equipment and lights (15). As illustrated in Figure 6.10, in the first three months, the electricity consumption amounts to over 6,000 wh per month and then decreases by 1500 wh to 2000 wh for the end of September. It then increases to around 6,000 wh for the last three months. These ups and downs align well with the climatic condition of Sydney and its summer–winter cycles. For the electricity consumption, there are other elements of area lights, miscellaneous equipment, vent fans, pump auxiliary and space heating that have minor impacts and do not experience remarkable variations.

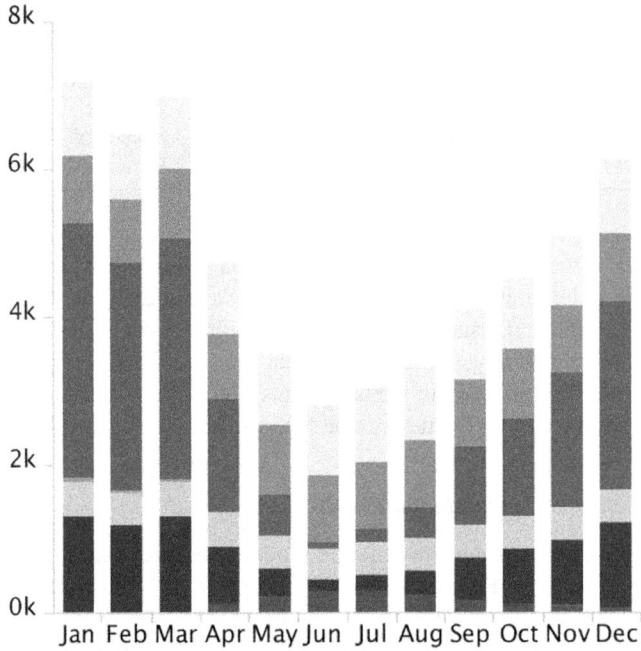

Figure 6.10 Monthly Electricity Consumption (wh) for Baseline Model

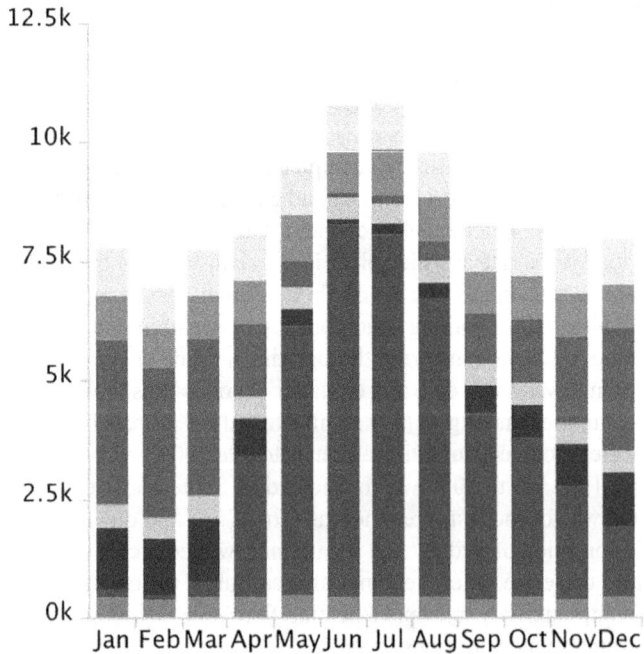

Figure 6.11 Monthly Total Energy Consumption (wh) for Baseline Model

For the total monthly energy consumption, as depicted in Figure 6.11, the peak is recorded at over 10,000 wh in the months of June and July. The significant element of total energy consumption is the space heating load, which is the amount of energy required for maintaining the space in the comfort band and includes both latent and sensible loads. The plotted graph in Figure 6.11 experiences a stable trend from January to April at the amount of 7500 wh and then increases gradually to reach the peak. This consumption diminishes in the last four months. In the total energy consumption calculation, in addition to the heating and cooling loads, the other elements of area lights, miscellaneous equipment, vent fans, pump auxiliary and hot water have been again estimated as the same as monthly electricity consumption. Since these items are more relevant to the occupancy profile and their estimation is based on fixed assumptions (Table 6.2), their effects are not that considerable or influential.

6.3.4   Case Optimisation Procedure

Given the simulation results for the baseline case model, and according to the developed AI and BIM integration framework, the optimisation procedure started with identifying the model elements and their current specifications of the model (see Table 6.1). This task was done through making schedule queries of data and federating the data from multiple disciplines of the case, including architectural, structural and mechanical elements. The reason for such federation is that the targeted 13 variables of optimisation belong to these case disciplines. The data schedules were then propagated via IFC and proceeded with Revit DBLink application in order to synthesise the data into Microsoft Access and Excel formats, readable by the external software of Matlab (Figure 6.12).

Upon making the database accessible by Matlab, the entity classifications of the database could be interpreted by the Matlab integrated simulator as the target variables. The continuous and categorical parameters were then differentiated and operated by ANN and DT algorithms under the hybrid objective function. By operation of the algorithms, a preliminary energy estimation resultant from the hybrid objective function was aggregated and minimised by the GA optimisation algorithm in order to reveal the optimised values for the target variables, as illustrated in Figure 6.13.

6.3.5   Case Optimisation Results

The optimised values of the model are found in Table 6.3. As it can be seen, there are significant changes in the model elements of the target variables. External wall material was set to a wood framed wall with a U-value of 0.54 W/m^2.K and 1.2 cm air film as the insulation layer. Roof, floor and window glazing were optimised to the metal roof, RC floor and triple-glazed (for all orientations) with U-values of 0.388, 0.199 and 0.66, respectively. The model was further oriented with 20 degrees toward the north, which is evidently the most suitable orientation for mild and temperate climates like Sydney by providing the daytime heating and cool sleeping required (14).

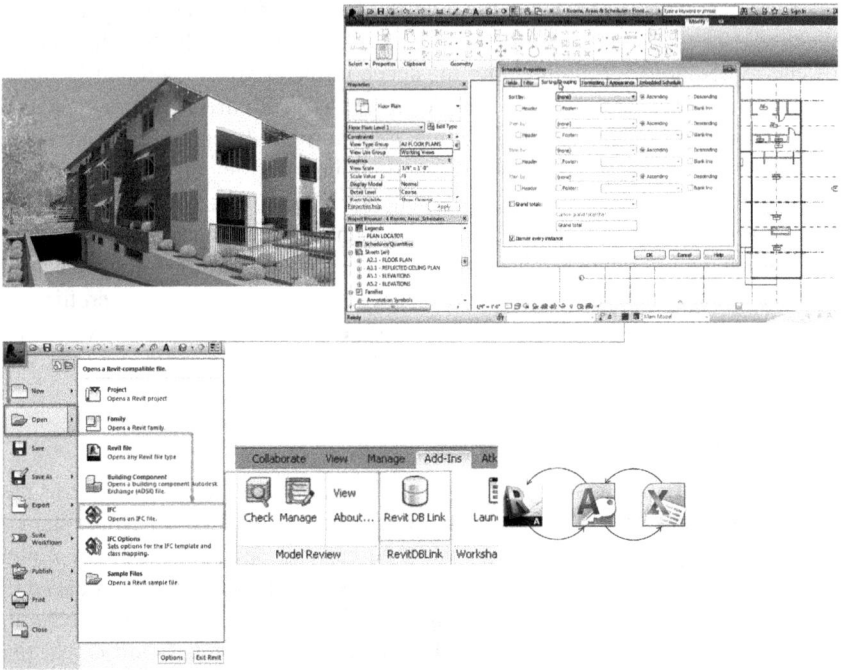

Figure 6.12 Database Development and Exchange Processes of the Case Study

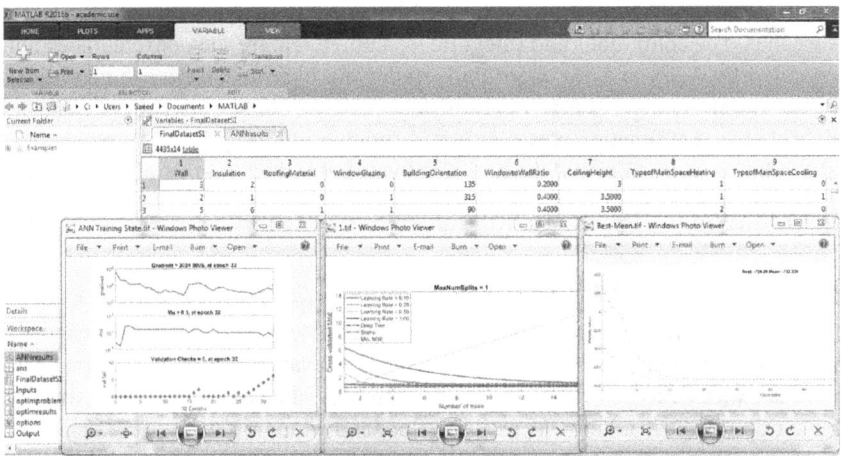

Figure 6.13 Matlab Interface during the Operation Process

Table 6.3 Optimised Baseline Construction Specifications

Construction Component	Specification	U-Value (W/m².K)
External Wall	ASHRAE 90.1–2007 Wood Framed Wall	0.54
Insulation	Air Film	0.209
Roof	ASHRAE 90.1–2007 Roof Metal	0.388
Floor	ASHRAE 189.1–2009 Floor	0.199
Windows	Triple-Glazed	0.66
Orientation	240°	
Ceiling Height	2700 mm	
Area of Rooms Cooled	718 m²	
Area of Rooms Heated	729 m²	
Type of Main Space Heating	Steam/Hot Water System	
Type of Main Space Cooling	Window/Wall Unit	
Lighting	4.16 W/m²	
Window to Wall Ratio	38%	

This orientation receives the passive heating of living areas during the day and cooler, southerly sleeping areas. In winter, the wood framed wall separates the zones and transfers solar warmth to sleeping areas since it has poor thermal mass and so has the higher level of thermal transfer. In summer, passively shaded clerestory windows along the spine would allow hot air to escape from bedrooms in summer while allowing in a small amount of winter sun. The ceiling height decreased to 2.7 metres; however, the areas of rooms cooled and heated were not considerably changed. The reason behind this observation may be due to the intensity of the other components' optimisation and the lower capability of the current BIM application in changing the layout of the design. The current BIM is fully powerful with potential in tweaking building envelop and physical properties, HVAC and, as a whole, geometrical and semantic attributes. Nevertheless, it lags behind in connecting building layout with the topological attributes (16). Furthermore, types of main space heating and cooling were transformed to the steam/hot water system and window/wall unit, respectively. Lighting intensity decreased to 4.16 W/m² but window to wall ratio increased to 38%.

Updating the baseline model with the resulting specifications from the optimisation process, the energy simulation was again conducted for obtaining the energy estimation by Revit-GBS simulation. Figures 6.14 and 6.15 indicate the monthly electricity and total energy consumption for the optimised model, respectively. Figure 6.14 shows that the electricity energy consumption has almost reduced by half and reached to its minimum level from April to December. In the first three months, the electricity consumption is around 3,000 wh per month and then decreases by 500 wh till the end of year. For total energy consumption, as indicated in Figure 6.15, although the same peak is occurred again around 10,000 wh in the months of June and July, the amount of the energy consumption for the other months was reduced significantly as well. The graph of the monthly total energy consumption (Figure 6.15) indicates the monotonous trend from January to April, with the amount of 3000 wh

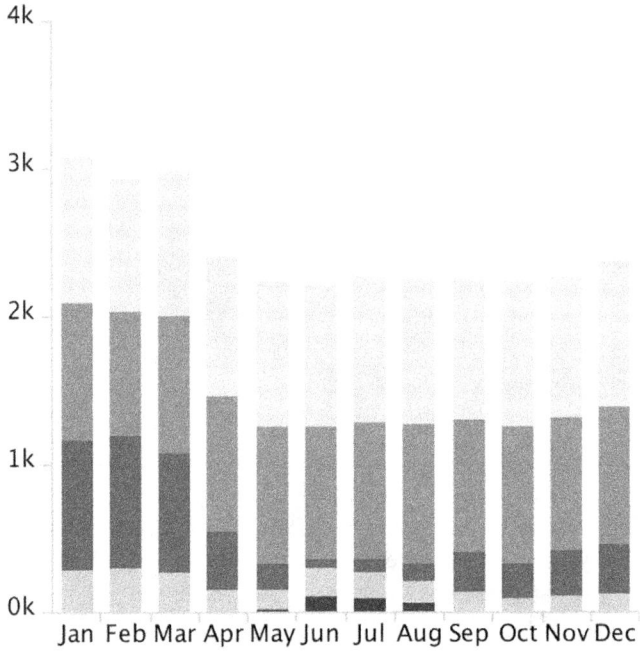

Figure 6.14 Monthly Electricity Consumption (wh) for Optimised Model

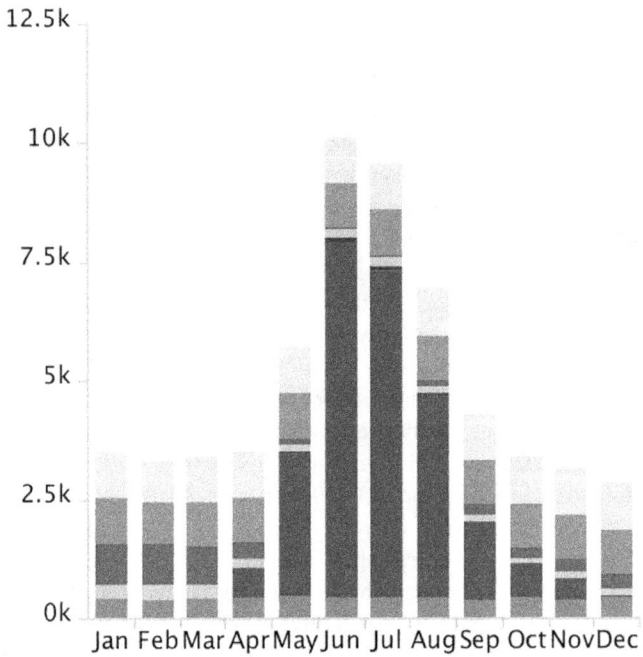

Figure 6.15 Monthly Total Energy Consumption (wh) for Optimised Model

(less than half of the baseline model performance). It then rises to the maximum in June while the peak is diminished again to the least by getting to the end.

Visually comparing the energy consumption graphs of the optimised model (Figures 6.14 and 6.15) with the baseline model (Figures 6.10 and 6.11), it can be stated that the electricity consumption has been significantly reduced, while the trend of total energy consumption is to some extent similar. This observation could be because of the fact that the optimisation was more focused toward the parameters that have direct effect on the electricity consumption aspects of residential buildings. In fact, the target parameters of energy optimisation resulting from the Delphi study are more inclined to the electricity energy plan rather than the other aspects like fuel consumption. Routinely, the other elements of energy consumption, such as area lights, miscellaneous equipment, vent fans, pump auxiliary and hot water, were also considered in the simulation.

For providing a holistic view on the validation procedure and results thus far, Figure 6.16 summarises the steps and outcomes of the case study. This figure presents five consecutive steps of using the case study, considering the simulation parameters, BIM simulation results for the existing and baseline model, framework operation and energy optimised results. The critical simulation parameters were categorised into the project information, building program, model elements, project variables and climatic condition while LoD 300 and the design stage were twice scrutinized. Progressively, three crucial outcomes of the annual electricity consumption, annual fuel consumption (gas) and annual peak demand were obtained (aggregation of the monthly estimations for the baseline model) with the quantities of 58,001 kWh, 164,195 MJ and 15.3 kW, respectively. These figures are within the range of the average household energy consumption released by Australian Energy Regulator in March 2015 (17).

Through the next step, the BIM-inherited EED integration framework was activated, and AI-enabled BIM was operated by three major operators of database development, database exchange and database optimisation, and the energy estimations were then delivered for the optimised model. As mentioned, a significant decline can be observed in the energy consumption level of the optimised model in which the annual electricity consumption has been almost cut by half, reaching to 29,531 kWh. Such a finding is more corroborated by the remarkable decline in annual fuel consumption (gas) and annual peak demand by recording 109,076 MJ and 14.0 kW, respectively (Figure 6.16). According to the Australian Green Building Council, these figures are within the range of the highest ratings for the energy category of residential buildings, five to six stars, audited by this body (18), which puts more highlight on the reliability of the framework.

6.3.6 Optimisation Reliability Tests

The last section of the validation and verification stage tests the result-oriented reliability of the optimisation runs and the significance of the results delivered by the framework. The optimisation engine used in the framework is GA, which has impeccable performance in simultaneously exploring and exploiting the solution

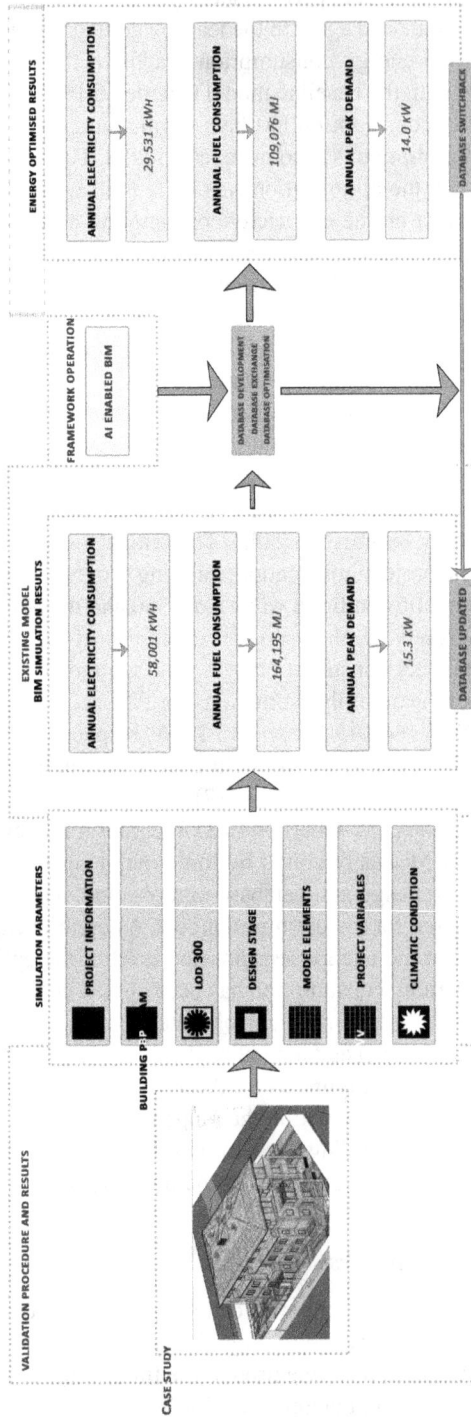

Figure 6.16 Validation Procedure and Results

realm. A growing number of theoretical justifications and successful applications to real-world problems confirm GA as a robust and decent optimisation technique. However, as to its heuristic nature, it could be paramount to finding the threshold of the approximate performance guarantee for the algorithm (19). The goal here is to find a circled ratio of the minimum and maximum values for the optimisation problem that the algorithm is supposed to handle in terms of the feasibility and reliability. In other words, in this method, an acceptable threshold should be first set for the lower and upper boundaries of an optimisation problem. In the second step, the result of the optimisation should be investigated to realise whether the result falls to the boundary or not.

Therefore, the solution space, the generations that the GA algorithm produced as the energy estimations based on the input variables, was first converted into a first-order approximation hyper-surface (p) to create its circular boundary. A reliability threshold should then be set as the constraints to determine the boundary for the solutions space ($\beta_{i,j}$). To do so, the Australian Green Building Council guideline on the minimum and maximum of energy savings from 4 to 6 Green Star rating, amounting to 42% to 72%, was applied as the reliability threshold (18). In this step, the optimisation result should correspond to the reliability threshold, which makes the first-order approximation of $p = \theta \ (\beta_{i,j})$. Figure 6.17 hypothetically

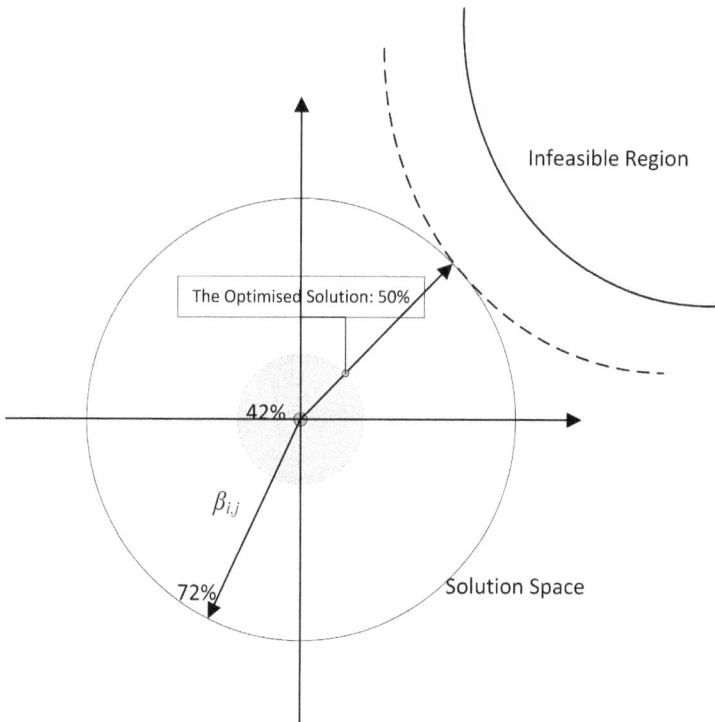

Figure 6.17 Optimisation Reliability Test

illustrates this approach to depict the probabilistic constraint in the solution space (U-space), the corresponding constraints of 42% and 72% and the feasible region. Hence, minimisation of the solutions space from 58,001 kWh to 29,531 kWh was found to be encompassed within the boundary and considered to be satisfactory.

The second test is to prove that the rising gap between the results of the baseline and optimised models is statistically significant. Since there are two sets of quantitative results for the same procedure of the energy estimation, a 2-paired sample T-Test was selected (20). The underlying concept is simply to calculate the average difference between the measurements. Then, determine whether there is statistical evidence that the mean difference between paired observations is significantly different from zero or not. The calculation formulae are as follows:

$$t = \frac{\overline{x}_{diff} - 0}{s_{\overline{z}}} \tag{6.1}$$

$$s_{\overline{x}} = \frac{s_{diff}}{\sqrt{N}}, \tag{6.2}$$

where X_{diff} is the sample mean of the differences, N denotes the sample size (the number of compared cases), S_{diff} indicates the ample standard deviation of the differences and S_x refers to the estimated standard error of the mean (s/sqrt(n)). The calculated t value is then compared to the critical t value with $df = N - 1$ from the t distribution table for a chosen confidence level. If the calculated t value is greater than the critical t value, it is concluded that the means are significantly different. Accordingly, the resulting quantities of the annual electricity and fuel consumption along with the peak demand of the case study were considered for the test. Table 6.4 indicates the calculation outcomes for these data. Evidently, there are three paired data for the baseline and optimised models (N = 3), which results in 1.499 as the t value and 9 as the standard deviation (df). This table shows that the results of the baseline and optimised models are significantly different from zero by 0.168 at 95% of confidence interval, which further verifies the statistical significance of the optimisation performance and validates the applicability of the framework.

Table 6.4 Paired Sample T-Test Calculations

Model/Observations	Annual Electricity Consumption (kWh)	Annual Fuel Consumption (MJ)	Annual Peak Demand (kW)
Baseline	58,001	164,195	**15.3**
Optimised	29,531	109,076	**14**
N (Number Of Samples)		3	
t Value		1.499	
df (Standard Deviation)		9	
Statistical Significance		0.168	

6.4 Sensitivity Analysis

Sensitivity analysis is an effective method to find out how mathematical models, machine learning packages or resultant frameworks react to a range of uncertainties. It is a technical procedure over the re-calculation of the outcomes in view of various assumptions made to assess the impact of independent variables on the dependent factor (21). Sensitivity analysis can be applied for different purposes, such as evaluating the validity of the results, better view toward the relationship between inputs and outputs, identifying model errors and reducing uncertainty via focusing on the significant factors (22). Therefore, in this section, a sensitivity analysis test was run to understand how sensitive the framework is and how some possible uncertainties impact on the framework, including missing data and noise interferences.

This approach started with establishing the framework and its dataset in Matlab and setting rules of data manipulations. According to Yeung, Cloete (23), machine learning-based models could be approximated with regression batches so that their sensitivity can be assessed through regression performance criteria such as MSE, R and MAE, as introduced in Chapter 5. Furthermore, it is recommended to consider random data eliminations of 10%, 20%, 30% and 50% from all variables involved in cases of large datasets. Moreover, data noises were involved via Matlab's automatic noise toolbox to model possible errors of regression approximation. Hence, these rules were capitalised on the platform and performed to activate regression analysis and the outcomes record. Full dataset regression was first run to create the baseline model leading to be compared with four uncertainty scenarios. Figure 6.18 indicates the data records and regression test for the full dataset are its baseline model. Analysing the current strength of the baseline regression in Table 6.5, MSE, R and MAE records of 1020, 0.76 and 776.18, respectively, confirm a significant correlation with the full dataset performance.

Subsequently, four sensitivity analysis scenarios were conducted to assist with the comparative evaluation of regression values and to reveal any significant discrepancies arising from exposure to uncertainties. Figure 6.19 illustrates the regression tests that were run for the 10%, 20%, 30% and 50% scenarios. As mentioned throughout this section, these scenarios were composed from randomly eliminating data from the dataset of 13 variables along with the automatic noise interference to the data to identify possible impacts of regression approximation.

Schematically, the plots show a good performance of agreement with a steady trend line. In spite of gradually random dataset reductions from complete to 50%, the overall trends of the regressions were almost kept fixed and reliable. Analytically, Table 6.5 indicates the figures of the regression performance criteria for uncertainty scenarios. It can be inferred from this table that the framework performs well and maintains homogeneity and consistency up to 30% of data reductions and noise interferences with the R values of 0.76, 0.75 and 0.75 for 10%, 20% and 30% scenarios, respectively. However, the sensitivity of

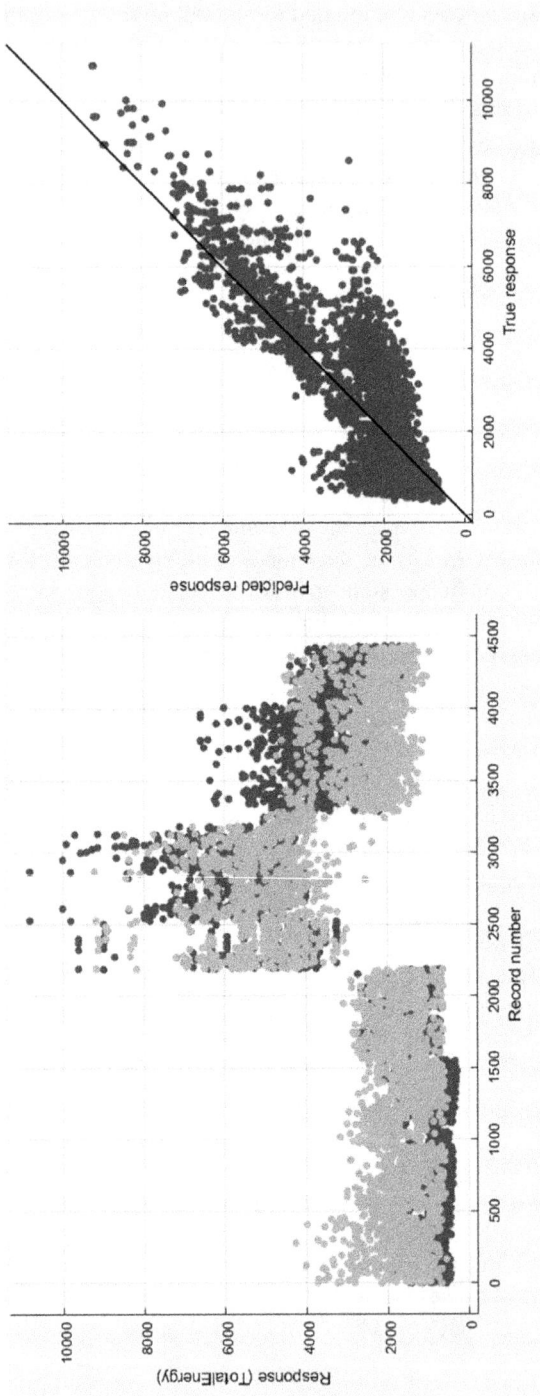

Figure 6.18 Regression of the Baseline Model

Table 6.5 Regression Results of Baseline vs. Sensitivity Analysis Scenarios

	Baseline Model	*10%*	*20%*	*30%*	*50%*
R	0.76	0.76	0.75	0.75	0.72
MSE	1020	1032	1043	1042	1060
MAE	776.18	779	795	798	803

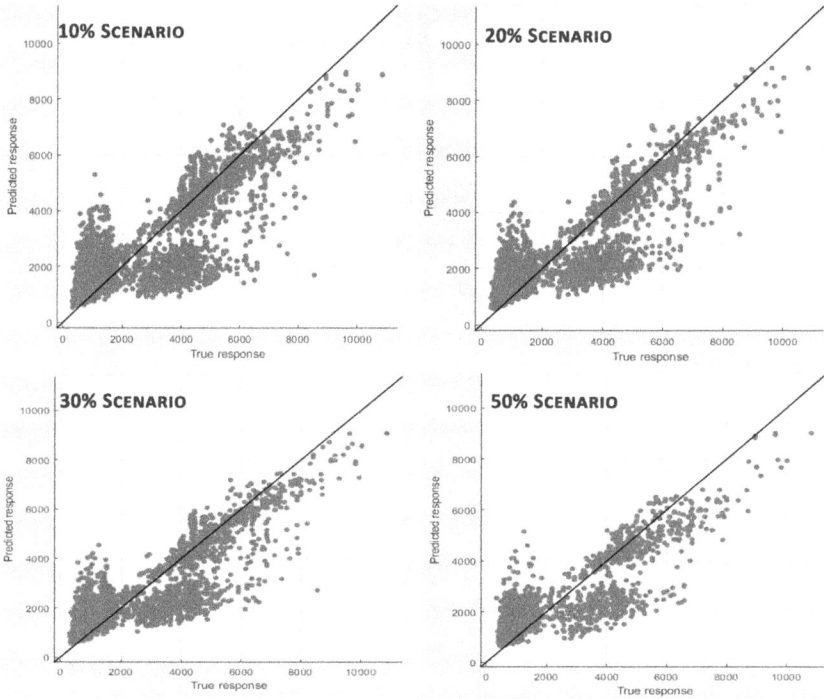

Figure 6.19 Regression Sensitivity of Different Scenarios

the framework is a bit intensified in the 50% scenario due to the very extreme uncertainty that is applicable for very large datasets and few number of parameters. The R value slightly dropped to 0.72 and MSE and MAE increased to 1060 and 803.

All in all, considering the conducted sensitivity analyses, it can be mentioned that the framework is valid and secure against the uncertainties of random data missing and noises up to 50% of its database asset. This achievement is superiorly remarkable compared to similar studies in the literature (22, 24).

6.5   Summary

This chapter presented the last part of the analysis and results of this research on four major sections: optimisation process, BIM-inherited EED framework development, testing and validation of the framework and sensitivity analysis. It was revealed that GA algorithm performed well on the hybrid objective function and converged into the optimum solution, quickly and properly. This achievement finalised the AI algorithms stage and grounded a steppingstone for developing the framework of AI-enabled active BIM. The database development, database optimisation, database exchange, database switchback and database update constituted five key elements of the BIM-inherited EED framework, which were established in series and processed integrally. This framework was tested through using a real residential building located in Sydney and via running a comparative energy simulation pre- and post-framework application (the baseline and optimised cases). The outcomes indicated around 50% reduction in the electricity energy consumption and 66% saving in the annual fuel consumption of the case study. This performance was verified by an optimisation reliability test and paired sample T-Test for statistical significance. The framework was further analysed against different uncertainty scenarios and found to be secured and reliable for the data reduction and noise interferences of up to 50%.

Note

1   This case study is different from the building case which was used for generating energy simulation datasets in Section 6.1.

References

1   Tuhus-Dubrow D, Krarti M. Genetic-algorithm based approach to optimize building envelope design for residential buildings. Building and Environment. 2010;45(7):1574–81.

2   Cohen PR, Feigenbaum EA. The handbook of artificial intelligence: Butterworth-Heinemann; 2014.

3   Sarker RA, Newton CS. Optimization modelling: A practical approach: CRC Press; 2007.

4   Tao F, Zhang L, Laili Y. Configurable intelligent optimization algorithm: Springer; 2014.

5   Sumathi S, Paneerselvam S. Computational intelligence paradigms: Theory & applications using MATLAB: CRC Press; 2010.

6   Rao SS, Rao S. Engineering optimization: Theory and practice: John Wiley & Sons; 2009.

7   Kensek K, Noble D. Building information modeling: BIM in current and future practice: John Wiley & Sons; 2014.

8   Mordue S, Swaddle P, Philp D. Building information modeling for dummies: John Wiley & Sons; 2015.

9   Teicholz P, Sacks R, Liston K. BIM handbook: A guide to building information modeling for owners, managers, designers, engineers, and contractors: Wiley; 2011.

10 NATSPEC. BIM and LoD: Building information modeling and level of development. Construction Information System; 2013.

11 Yang H-C, Shiau Y-C, Lo Y-H, editors. Development of a BIM green building evaluation software. Applied System Innovation (ICASI), 2017 International Conference on, IEEE; 2017.

12 CIBSE GA. Environmental design. The Chartered Institution of Building Services Engineers, London; 2006.

13 World meteorological organization [Internet]; 2016. Available from: www.wmo.int/datastat/wmodata_en.html.

14 Bambrook SM, Sproul AB, Jacob D. Design optimisation for a low energy home in Sydney. Energy and Buildings. 2011;43(7):1702–11.

15 Sørensen B. Life-cycle analysis of energy systems: From methodology to applications: Royal Society of Chemistry; 2011.

16 Gerrish T, Ruikar K, Cook M, Johnson M, Phillip M. 6 BIM in practise: Industry case studies. Advances in Construction ICT and e-Business; 2017.

17 Regulator AE. Guidance on electricity consumption benchmarks on residential customers' bills, Australian Energy Regulator, Melbourne, Australia; 2015.

18 GBCA. The value of green star: A decade of environmental benefits. Green Building Council of Australia, Melbourne, Australia; 2013.

19 Williamson DP, Shmoys DB. The design of approximation algorithms: Cambridge University Press; 2011.

20 Corder GW, Foreman DI. Nonparametric statistics for non-statisticians: A step-by-step approach: John Wiley & Sons; 2009.

21 Hopfe CJ. Uncertainty and sensitivity analysis in building performance simulation for decision support and design optimization. PhD diss, Eindhoven University; 2009.

22 Tian W. A review of sensitivity analysis methods in building energy analysis. Renewable and Sustainable Energy Reviews. 2013;20:411–19.

23 Yeung DS, Cloete I, Shi D, Ng WY. Sensitivity analysis for neural networks: Springer; 2010.

24 Nourani V, Fard MS. Sensitivity analysis of the artificial neural network outputs in simulation of the evaporation process at different climatologic regimes. Advances in Engineering Software. 2012;47(1):127–46.

7 Conclusion

7.1 Review of Background, Problem, Aim and Method

GB is being promoted in the construction industry in order to minimise detrimental impacts of the massive upsurge of construction activities, resource depletion, waste generation, energy consumption, carbon emission etc. (1). It entails integrating principles of sustainability into building design and construction activities throughout the whole lifecycle of a building facility. This can include every actor and action involved in a project being committed and responsible for undertaking sustainable practices (2). Despite this strategic focus, challenges such as a lack of whole system thinking, construction process complexity, lack of communication and coordination among parties involved, unavailability of innovative approaches and tools and lack of integrated design methods hamper the proposed shift toward GB practices (3). The problem of translating these strategic objectives into concrete actions at the EED level has in particular remained unresolved (4).

Therefore, the construction industry has attempted to tackle a wide range of challenges and their unsustainable impacts, such as excessive energy consumption, by harnessing the capabilities of BIM (5). Of these, connecting the often-fragmented processes of a building design through automated, integrated and streamlined communication has remained the main application, while efficiency gains with digitised and standardised energy simulation procedures are another benefit (6). The BIM process has the potential of offering numerous advantages in energy consumption estimation. It represents the building as an integrated database of coordinated information and can expedite the EED process through providing the opportunity for testing and assessing different design alternatives and materials' impacts on building energy consumption (7). In spite of such great potentials, EED still lags in fully embracing BIM (8).

The passiveness of BIM in endorsing intelligent decision-making platforms, the lack of ideal interoperability between BIM and energy analysis software packages, the dearth of inbuilt practices of calculative, predictive, simulative and optimisation methods of EED in BIM and the need for more inputs and technical specifications in improving the semantic-rich content of BIM models hinder the complete diffusion of BIM into EED (BIM-EED). This premise has triggered a new research direction, known as the introduction, inclusion and integration of AI

DOI: 10.1201/9781003207658-7

into BIM-EED. The research element of this book focuses on developing, facilitating and optimising EED in an integrated manner by means of BIM and AI. It represents an alternative solution for building design and construction projects to embrace sustainability. This assertion seems a workable way of introducing BIM and AI to EED. However, very little is known about the methods through which this could be achieved. In essence, research on the interface between BIM-EED and AI is very limited. In particular, not much in the way of an explicit body of knowledge from the built environment context has been developed to investigate the integrated effects of AI and BIM on EED.

AI facilitates access to transparent information, enhances accountability, increases accuracy of data, enables automation and integration of activities in BIM, elevates the level of agility and generates insights for better decision-making and optimisation of EED processes. It handles the problem of interoperability by shifting the current post-design routine of energy analysis and optimisation in using the external software into the inbuilt EED in BIM. AI further drives 'what if' scenario analysis for choosing various design parameters by applying prediction, classification and optimisation algorithms on the operational energy of building model in the design stage. Moreover, the issue of consistency and homogeneity of the EED data can be mitigated, based on the integrated platforms of AI and BIM. Therefore, this book aimed to develop an AI-based algorithm to optimise energy efficiency at an early design stage in BIM. It is to obtain an initial estimate of energy consumption of residential buildings and optimise the estimated value through recommending changes in design elements and variables. This aim was pursued through the below objectives:

1 Examining the potential and challenges of BIM to optimise energy efficiency in residential buildings
2 Identifying variables that play key roles in energy consumption of residential buildings
3 Investigating AI-based algorithms in energy optimisation
4 Developing a framework of AI application in BIM in terms of energy optimisation purposes and processes
5 Assessing and validating the functionality of the framework using case studies

In order to achieve the stated aim and objectives, a mixed method approach following a design of *QUAL → QUAN* was designated for this study, which entailed conducting a preliminary qualitative data collection method to serve the subsequent quantitative phase. This sequential exploratory approach started with the application of qualitative instruments to the exploratory tasks of objectives one and two. It included examination of the potential and challenges of BIM and EED and identification of the significant variables in the energy consumption of residential buildings via literature study and Delphi approach. It was then factualised with quantitative instruments to address the third to the fifth objectives, where the simulation method was used to generate the building energy datasets and simulate

AI algorithms to investigate their functionality for energy optimisation. It was then followed with developing the integration framework of AI and BIM and, finally, validating the framework through case study verification.

7.2   Review of Research Processes and Findings

As mentioned, the aim of this study was to develop an AI-based algorithm to optimise energy efficiency at an early design stage in active BIM in order to obtain an initial estimate of energy consumption of residential buildings and optimise the estimated value through recommending changes in design elements and variables. Hence, the key processes and findings of the research aim for this book that flow through conducting its five research objectives are discussed in the following.

7.2.1   Objective 1: Examining the Potential and Challenges of BIM to Optimise Energy Efficiency in Residential Buildings

This objective was primarily fulfilled through reviewing an overarching theoretical framework of three theories of sustainability, information and optimisation. It was revealed that the definition by Brundtland (9), 'development that meets the needs of the present without compromising the ability of future generations to meet their own needs', formed the basis for this book. So, developing knowledge at interdisciplinary levels and efforts of modelling and evaluating sustainability were among the actions that could positively contribute toward this definition. This procedure is a cycle of endless development and adjustment that indicated the interaction of sustainability, information and optimisation theories require continued supervision and analysis for finding an intelligent decision based on CDE. It was disclosed that the cutting-edge scientific discoveries in information theory together with computer-aided algorithms in optimisation theory can facilitate managing the scale and spectrum of sustainability issues and allow for the smooth evolution of data.

It was then presented that information theory and its functional driver, informatics, works on the information process toward sustainability. This momentum can be significantly conducted when CDE as the single source of information could be applied to drive BS1192 information management implementation. Via this virtue, this batch of information is optimised through optimisation theory paradigms via setting objective function, mixing continuous and integer variables and convex function for reaching the maximised effectiveness of sustainability. Therefore, such a theoretical integration sets the journey to the downstream and practical level that should be elaborated in the construction field. However, it was realised that these concepts are mostly applied in isolated case in the construction industry and that an interdisciplinary attempt should be made to build an inextricable link toward their functional approaches for addressing the aim of this study.

Hence, sustainability was narrowed into sustainable construction drivers. BIM, then, came to the fore as the representative of informatics, CDE and BS1192 applications in the construction industry to store, transmit and process the optimised information. The optimised information is that which is optimised through AI as the practical machine of optimisation theory paradigms to convex them. Through this setting, the integration of three fundamental theories of sustainability, information and optimisation was built on CDE, linked to their three functional drivers of sustainable construction, BIM and AI and established as a theoretical framework for the aim of this book.

It was also identified that there are two important concepts, passive and active BIM, which are opposed to each other. Passive BIM lacks intelligent decision-making and cannot sufficiently provide sustainability analysis data. However, active BIM is AI-enabled BIM, which supports the intelligent decision-making frameworks in its structure and contributes to sustainable construction by streamlining design optimisation capabilities. This fact consequently underlined trilateral interactions of sustainable construction drivers, BIM and AI by discussing the calculative, predictive, simulative and optimisation methods in the common ground of EED.

Performing Objective 1 was followed by a comprehensive review on BIM-EED in light of BIM diffusion stages into EED. The available literature on the topic was identified and verified through a systematic review; BIM-EED was categorised into three stages of BIM-compatible, BIM-integrated and BIM-inherited EEDs; and their state of the art was reviewed. In principle, BIM-compatible EED referred to the first level of BIM adoption and indicated the central role of data exchange between BIM and a first generation of energy simulation software. One step ahead, BIM-integrated EED underscored the direct integration of BIM and a second generation of energy applications as to the maturity of the intelligence growth from data to the information level. The third level of BIM diffusion into the EED, BIM-inherited EED, denoted the internal interoperability and inbuilt function of the third generation of energy simulation in the homogeneous BIM platform.

It was indicated that although BIM-inherited EED constitutes the least share of the applications so far, there is a shift from lower to higher levels of BIM implementation in EED. This observation is also imperative for a move to the upper levels of interoperability and LoD in this area. Thematic analysis showed that case study-based analytical, tool/prototype development and conceptual studies are in the first to the third rank of target for researchers. Hence, it was concluded that conceptual frameworks should be validated through case studies in order to generate more reliable prototypes of BIM-EED. Gaps were highlighted according to three major categories: confusion, neglect and application in the extant literature.

Confusion spotting indicated a gap in interpreting BIM as a process vs. as a tool. This observation was corroborated by competing interpretations due to the BIM-EED adoption confusion. The second category, neglect, was found to be the biggest part of gap spots where it included three subgroups: overlooked, under-researched and lack of empirical support. Analysing this gap demonstrated that interoperability and LoD are mostly overlooked. Furthermore, AI, machine learning and data analytics are under-researched areas in BIM-EED, and the majority of conceptual studies suffer from a lack of empirical support via proof of concept. Finally, it was concluded that BIM-EED are overwhelmed with the application gap, focusing on the sole application of energy simulation and the dominance of a technology-oriented view toward BIM.

Conducting Objective 1 resulted in significant achievements and implications for application in this book. It was recognised that considering the three theories of sustainability, information and optimisation and their integration path narrowed to EED, BIM and AI lead to BIM-inherited EED, as the active BIM and based on CDE concept. It was found to be the particular focus for this research and framework development objective. BIM-inherited EED was attributed to the highest level of BIM adoption into EED, and hence, in order to achieve the most out of BIM potential, the AI-based active BIM was developed on the BIM-EED basis. The parametric definition of BIM was further substantiated through AI algorithms to enable data analytics and energy optimisation inbuilt in BIM.

The thematic and research outcomes analysis of the literature showed that an established approach should be designed with BIM-EED research in order to develop a conceptual framework and verify its functionality. Therefore, the study was designated to comprise a consolidated combination of the theoretical background, simulation-based framework development and framework validation via the case study. In addition, the scrutiny of BIM-inherited EED applications determined the package of Autodesk GBS-Revit® to analyse and test the validity and reliability of the framework. Finally, to enhance the precision, depth and inclusion of BIM models in the framework, *LoD 300* was found to be target for the framework development and verification goals as the topmost LoD for the design stage.

7.2.2   *Objective 2: Identifying Variables That Play Key Roles in Energy Consumption of Residential Buildings*

Performing Objective 2 commenced with a review of the design and construction classifications of parameters affecting buildings' energy consumption, including physical properties and building envelop, building layout, occupant behaviour and HVAC and appliances. It was inferred that physical properties and building envelop are the key drivers in thermal losses and are vital for optimisation in the early design stage. This category includes the building configuration, shell and material-related elements including walls, insulation, floors, roofs and windows representing thermo-physical properties and the mechanisms of heat transfers or heat stores into the geometries. It was also identified that the building layout category of variables mostly covers the design brief of the building, such as the number of rooms or the number of spaces requiring heating and cooling, size of the building, architectural shape and building orientation. It was found out that it is a bit difficult to generalise or quantify the complex interrelationship of the planning and layout of spaces on energy consumption requirements.

Occupant behaviour was further recognised as one of the most important but challenging categories in influencing building energy consumption. It was found difficult to investigate, especially in the design stage, due to complicated characteristics and unpredictable personal behaviour. It was also revealed that a clear understanding of the schedule of building operation is important to the overall accuracy of the energy estimation. Last but not least, HVAC and appliances contain the parameters relevant to heating, ventilation, air conditioning and electrical appliances within buildings. It turned out that this category is very significant in optimisation since electricity has the highest percentage in HVAC among all building services, installations and electric appliances. It was also confirmed that modelling, simulating and optimising these variables are complicated because each appliance and device has a different production configuration with different energy efficiency index.

A three-round Delphi study was designated for complementing the literature and brainstorming of experts for building energy variables and their implications toward the design stage, BIM and optimisation, prioritisation and confirmation of the variables. In the first round, 35 variables were extracted from the qualitative answers through a quick textual analysis. Furthermore, with respect to the open-ended question in the first round on the respondents' implications for the applicability of variables for BIM, optimisation and design stage, very insightful achievements were obtained. In terms of BIM compatibility, it was stated by the experts that very generic variables cannot be considered in this study since the semantic values of very generic variables are not associated with the topological relationship within the BIM model. Therefore, the result of literature and Delphi was synthesised, in which a generic parameter of 'material' was divided into 'internal wall material', 'external wall material' and 'roofing material', falling into the predefined ranges of BIM families.

In addition, some of the experts mentioned that variables with two-fold effects also need to be omitted because BIM has not yet been equipped with intelligent fuzzy rules to identify the root causes of two-fold parameters, and so, these variables cannot be optimised. Therefore, the variables of 'building size' and HVAC system were broken down into variables of 'metres squared of rooms heated' and 'metres squared of rooms cooled', and the 'type of main space heating equipment used' and 'type of main space cooling equipment used', respectively. Checking the subjectivity and objectivity of variables was another recommendation by respondents, and as a result, a too subjective variable, 'morphology', was omitted due to still not being parameterised and measured in BIM. The respondents put also forward their implications regarding the design stage as the scope of this book and stated that the variables relevant to the occupancy profile of buildings cannot be involved in the design stage focused study as these variables are within the operation phase of a building project lifecycle.

Thus, applying technical recommendations from the respondents and running a normative assessment method via considering the 50% cut-off criterion rule resulted in 19 variables. A wide range of variables that covers different aspects of buildings' energy parameters were achieved by the end of Round 1. For instance, external wall materials, internal wall materials, wall thickness, insulation type, roofing materials, ground floor systems, window glazing types and door glazing types covered the physical properties and building envelop parameters. There were also parameters including ceiling height, metres squared of rooms heated, metres squared of rooms cooled, window to wall ratio and building orientation, which belong to the architectural design and building layout aspects of the influential parameters. Variables such as lighting, daylighting, external and internal shading addressed the required considerations in designing for energy efficient lighting and solar exposure. Finally, the parameters of cooling space and heating space equipment used applied the HVAC system design significance.

The prioritisation round (Round 2) was conducted with experts to prioritise the 19 variables derived from the first round and the 5-point Likert scale method. The cut of rule of receiving the Likert point equal to 3 or above resulted in having 13 variables with a Kendall's coefficient concordance of 0.149. The findings of this round indicated the emphasis of the respondents on the role of physical properties and building envelop in the energy consumption of residential buildings since the top three parameters of the ranking – insulation, roofing material and external wall material – belong to this category of variables. Moreover, the other parameters of window glazing type, ground floor system and lighting were also fixed in this round. With respect to the architectural design and shape, building layout pertinent variables including ceiling height, metres squared of rooms heated and cooled, window to wall ratio and building orientation were found significant in this round. Types of main space heating and cooling equipment affiliated with the HVAC category were also maintained. Finally, as a result of a prioritisation round, six variables – internal wall material, internal and external shadings, wall thickness, door glazing, and daylighting – were removed.

In the confirmation round (third round), 15 completed responses were obtained, and the Round 2 list was reconsidered in light of the consensus opinion of the respondents. Accordingly, the final matrix of 13 parameters and their relevant weightings along with the enhanced concordance score of 0.267 was attained. It was revealed that consistency was remarkably enhanced in the third round, where it attained 80% improvement in reliability and consistency of the results compared to the second round. These variables fixed the key asset for energy optimisation of buildings in the design stage through running AI (in the next objective) to exploit the parametric definition inherited in the BIM technology. These variables were then parametrised via AI approaches, including machine learning and data prediction, classification and optimisation algorithms in order to be approachable in the BIM environment through securing geometrical, topological and semantic associations.

7.2.3   Objective 3: Investigating the AI-Based Algorithms in Energy Optimisation

The third objective was intended to develop the package of AI algorithms for energy optimisation. Driven by the achievement of Objective 2, a full-scale parametric energy simulation was conducted on a hypothetical residential building as a case study and in four climates – temperate, tropical, cold and arid – to generate a dataset. The dataset size reduction techniques of metaheuristic-parametric approaches, including holistic cross-reference and evolutionary solver function, resulted in generating 4435 datasets including 13 inputs (the variables of Delphi) leading to the output (annual energy consumption). Trends and nature of the dataset for categorical and continuous variables were identified through descriptive statistics. It was indicated that, considering both continuous and categorical data, the generated output covered a wide range of distribution, which is of advantage for optimisation purposes so as to enable more precise optimisation.

Comprising both quantitative and qualitative parameters in one dataset led to developing both prediction and classification algorithms using ANN and DT, separately and integrally, in order to find the best type of inclusion of these algorithms. First, the ANN algorithm was established for the whole dataset with the conceptual architecture of 13 neurons of variables in the input layer, 7 neurons in the hidden layer and one neuron of annual energy load in the output layer. Different types of ANN training algorithms and percentage of training, testing and validating of data were tested, and the batch with 70%, 15% and 15% of the cases for the training, testing and validating of data under the Trainlm class were trained. ANN worked well on the data with a mean square error of 0.557228 and at the 26th iteration and the overall R value of 0.81186, which presented a strong performance as an objective function in the optimisation problem.

Second, the dataset including categorical and continuous parameters was again developed through the DT algorithm. The DT model was configured with containing the categorical data, integer classification of continuous parameters and annual energy load as the output with four levels – low, medium, high and excessive – of energy consumption. C4.5 training and the information-gain were calculated based on the entropy concept to apply learning the data and attribute selection. As a result, four types of tree structures – simple, medium, complex and bagged – were tested using the energy dataset. Confusion matrix plots and the comparative performance of actual vs. predicted were then employed as accuracy indicators. It was deduced that the bagged tree consisting of 90 complex trees significantly outperformed the other classifiers with an average accuracy of 87.30%.

Hybrid algorithm development was the third step in fulfilling this objective. The reason behind this matter was an attempt to develop a unified approach in handling both continuous and discrete parameters in their original format without need for transformation. Thus, the dataset was split into continuous and categorical groups, ANN and DT were run respectively on each dataset and the weighted average of the output was calculated. From the classification side, the bagged tree DT algorithm was trained on the data containing 15 complex trees in the structure and producing the acceptable error: 0.8 cross-validated error. From the prediction side, the ANN algorithm was established with 6, 7 and 1 neurons in the input, hidden and output layers and the best validation performance was recorded at the 109th iteration with MSE of 0.40974, which was satisfactory.

Therefore, the hybrid algorithm of prediction and classification was composed from coupling ANN and DT, which covered both quantitative and qualitative parameters with a hybrid error rate of 0.6. This rate of error was found to be satisfactory vis-à-vis that of similar studies. Moreover, the normalised predictive performance of a single ANN, single DT and hybrid model were plotted against the normalised actual energy data, and the superior performance of the hybrid model was confirmed.

This objective revealed that the hybrid algorithm of ANN-DT is capable of making a powerfully integrated engine for building energy optimisation in BIM.

7.2.4   Objective 4: Developing a Framework of AI Application in BIM in Terms of Energy Optimisation Purposes and Processes

The core contribution of this book was to achieve AI-enabled BIM-inherited EED, an integrated framework of AI application in active BIM, to optimise the EED of residential buildings. Hence, this objective was designed to include the achievements of Objectives 1, 2 and 3 and was conducted through, first, driving the optimisation engine and, second, performing the key stages toward framework establishment. For this reason, the developed hybrid algorithm and its included datasets (results from Objective 3) based on the identified variables in Objective 2 were identified as the fitness function and target parameters of optimisation. The GA algorithm was switched on and run on the objective function and data. Consequently, the algorithm reached superior convergence by 80 generations. Simultaneously, the methods of crossing over and mutation were applied to minimise the risk of early termination, and optimum diversity of the solutions was achieved via the average distance plot.

The framework development was composed of five main phases – database development, exchange, optimisation, switchback and updating. The primary purpose of performing these steps was to integrate AI algorithms of prediction, classification and optimisation with BIM and keep the homogeneity of the optimised data. Database development was mostly concerned with modelling and information standard strategies in the workflow. Representing model elements and associated technical components such as shape, size, location and orientation in the most accurate and realistic way was the prerequisite of the model federation of all disciplines involved in the BIM platform. However, beyond geometrical information, it was intended to include non-geometric and functional properties through adhering to LoD 300. It was found that this LoD headed further toward maintaining the semantic association with the geometrical and topological configurations of data enrichment of BIM models (results from Objective 1).

The database exchange was set to facilitate the exchange and interoperability of the developed database resulting from the database development. Two consecutive functions of IFC information export and its interoperability interface with Revit DBLink came to the fore to export and explore the data (results from Objective 1). More specifically, first, building layout, HVAC and physical properties and building envelop parameters were obtained as the project information (results from Objective 2). Entity class definitions were attached to the project information and sent for standardising and structuring according to IFC schema. The database was then synthesised with ODBC and made accessible for Matlab, and hence, the reciprocal relation between Matlab and BIM was achieved. Database optimisation gestated the optimisation procedure (results from Objective 3) of the exchanged database by exerting query-able ontologies on semantic, topological and geometric entities. Using parameter placeholders, the available values were specified in the Matlab integrated simulator by calling the AI package. The ANN-DT hybrid objective function was activated, and the initial prediction and classification results were delivered to GA in order to be optimised. Hence, the optimum combination of parameters was found and stored in a new database.

To ensure the dynamic integration of the optimised data with the existing RevitDB, the database switchback phase was executed. The key was to incorporate the control with the query command on the optimised database. As a result, this stage provided a system for BIM to link the modifications of the model database with the updated one. The last phase of the objective 4 was done for updating the database by overwriting the optimised parameters with their initial version. The object-oriented modelling concept was allowed to generate the comparative report of the BIM variations once the model updated. Furthermore, the automatic propagation of optimisation and relevant changes were expedited throughout the model. The performed phases were ultimately directed toward an AI-enabled BIM-inherited EED framework, which created the interdisciplinary data interoperability of EED in the seamless integration of BIM with AI algorithm packages.

7.2.5 Objective 5: Assessing and Validating the Functionality of the Framework Using Case Studies

The last objective of this book was to deliver a proof of concept for the developed framework and validate its performance toward an integrated optimisation of building EED. Henceforth, a real-life project of a multiunit residential building located in Sydney was used and modelled in Revit to validate the functionality of the framework. The parameters and constraints required for occupant behaviour and an energy simulation were determined and the baseline case was simulated in GBS in order to reveal the current energy consumption level of the building. As a result, monthly electricity and total energy consumption including electricity, hot water and gas for heat pump were calculated and the relevant graphs plotted. By aggregating these monthly consumptions, three crucial figures – the annual electricity consumption, annual fuel consumption (gas) and annual peak demand – were obtained with records of 58,001 kWh, 164,195 MJ and 15.3 kW, respectively. These quantities fall to the average household energy consumption in Australia, according to the authorities (10).

Upon implementing the AI-enabled BIM-inherited EED framework, significant changes were made in the model elements and the optimised construction specifications were recorded. The baseline model was updated with the optimised parameters and the energy simulation was again run to reflect the status of post-framework implementation. The outcomes of annual electricity consumption, annual energy consumption and annual peak demand were computed at 29,531 kWh, 109,076 MJ and 14.0 kW, respectively. These figures indicated a remarkable decline in annual electricity consumption, fuel consumption and peak demand from the baseline to the optimised case, especially for electricity energy, which was cut by half. The new quantities represented the sixth star class of energy consumption for the case study based on the Australian Green Building Council (10) standard and confirmed the practical function of the framework.

The validation objective continued with two more verification procedures to demonstrate the result-oriented reliability of the optimisation and framework operation. First, the concept of a reliability threshold was applied to position the energy optimisation results within the circled ratio of the minimum and maximum values for the optimisation problem. Therefore, a 4 to 6 Green Star rating of energy category of the GBCA (10) guideline, corresponding to 42% to 72% of energy consumption reduction, was set as the benchmark of reliability. It was approved that 50% of the decline in energy consumption resulted from applying the framework, perfectly matching with the benchmarking threshold. Furthermore, a paired sample T-Test was utilised to show the statistical significance between the baseline and optimised results. It was then specified that two pairs of data were significantly different from zero by 0.168 at 95% of confidence interval, which further verified the statistical significance of the optimisation performance and validated the applicability of the framework.

Finally, the sensitivity analysis test was run to identify the reliability of the framework in case of exposure to uncertainty and noise interferences. Therefore, four sensitivity analysis scenarios – of 10%, 20%, 30% and 50% of random data missing and noise interference – were conducted to assist with comparative evaluation of regression values and reveal any significant discrepancy arising from enforcing uncertainties. It was inferred that the framework works well and keeps homogeneity and reliability up to 30% of data missing and noise interference with R values of 0.76, 0.75 and 0.75 for 10%, 20% and 30% scenarios, respectively. However, the sensitivity of the framework was a bit increased in the 50% scenario, which is applicable for very large datasets with few numbers of parameters. Hence, the validity and security of the framework against four case studies of uncertainty scenarios including random data missing and noises up to 50% of the database asset were further justified.

Summarising the key processes and findings of the research objectives showed that Objectives 1 to 5, presented in Chapter 1, were fulfilled successfully and satisfactorily. Achieving these objectives has accomplished the aim of this book and contributed to the body of knowledge in different ways, as discussed next.

7.3   Contribution to Knowledge

Contribution to a body of knowledge is the most significant aspect of any research study (11). Bourke (12) itemised contribution to knowledge based on two criteria – originality and implications for practice. He decomposed originality into three subcriteria of conducting novel empirical works, carrying out research in an uncharted area and genuine synthesisation.

7.3.1   *Originality*

Thus, the originality of the contribution to knowledge was structured in terms of its subsets as the following.

- *Conducting novel empirical works.* Knowledge creation is usually processed through developing a new theory, improving an available theory and/or rebutting a theory, wholly or partially, in light of empirical data (13). In this study, novel empirical work was conducted on the basis of three broad theories of sustainability, information and optimisation and their practical aspects in the construction context: sustainable construction drivers, BIM and AI, toward the unified approach of AI-enabled active BIM-inherited EED (Objectives 1 and 2). A novel CDE approach was founded based on BS1192 to further generalise the applicability of this study for higher levels of BIM diffusion and implementation. The developed framework was established and validated via collecting empirical data (Objectives 3, 4 and 5). Evidence of originality for built environment research (11), this fact approved the transferability and applicability of existing theories from other disciplines to the built environment area.

*   *Carrying out research in an uncharted area.* Driven by collecting the empirical data, the research tried to authenticate AI-enabled active BIM-inherited EED in the built environment context. It was performed via linkage of the outcomes of Objectives 1 and 2 with the results from running Objectives 3 and 4. To the best of the authors' knowledge, not much research has been found applying BIM and AI integrally in the EED of the built environment field. In addition, the study attempted to apply previously developed theories of sustainability, information and optimisation from other disciplines of science, engineering and mathematics to a new area in the built environment context. This fact shows the efforts in confirming the validity of those theories in this field (11).
*   *Genuine synthesisation.* Objective 1 was primarily executed to identify the typologies of the book and the available categories of approach. Objective 2 was a synthesis of the literature review and experts' viewpoints on the building energy parameters and their significance in such studies. Objective 3 was also a synthesisation of previously separated methods of AI in prediction and classification algorithms. Ultimately, Objective 4 was specifically conducted to synthesise the outcomes of prior objectives into a unique framework development (AI-enabled BIM-inherited EED). Therefore, the rule of a genuine synthesisation – 'making a new interpretation of existing material' – was followed for evidencing originality (14).

7.3.2 *Implications for Practice*

Implications for practice (IFP) is the second category of expectation for research monographs in contributing toward practical knowledge. Such contributions are specifically significant for construction and the built environment due to the urgent need for enhancing the construction process efficiency (15). The IFPs of the present study's findings are presented next, drawing from the taxonomy proposed by Bartunek and Rynes (16), who mentioned potential audience, enhanced awareness and identification of new learning areas for IFP areas, which are used in the following discussion.

*   *Potential audience.* Drawing upon the practical application of an AI-enabled BIM-inherited EED framework in the design stage, architects, designers,

engineers, sustainability auditors and the research community can benefit as the first tier of audiences. In the second tier, BIM managers, design managers and project managers can take advantage of BIM diffusion levels in EED and set appropriate strategies in organisational implementation of those adoption stages.

- *Enhanced awareness.* As a result of the growing importance of sustainable design in the construction industry, this book will raise awareness of involved parties and potential audiences in the field. Identification of significant building energy variables and quantification of their level of significance, along with development of the framework that was verified based on the relevant industrial standard, can significantly enhance the knowledge and information of practitioners in putting them into practice. They can be informed from modern methods of EED, the challenges of prediction and optimisation of energy performance and BIM and AI effects on this procedure.

- *New learning areas.* Highlighting the practical challenges and issues arising from findings of a study is a source of learning and obtaining further knowledge for potential audiences. The findings of this book revealed that it is of crucial importance for designers and practitioners in streamlining EED in the BIM environment. It is done via obtaining an in-depth view of the BIM process, enhancing the LoD of design models and appropriately setting model elements with the topological, geometrical and semantic associations of BIM models. These challenges alongside the others, like most effective methods of BIM and AI integration, shed a light on areas necessitating further reading and education for the target groups of practitioners.

7.4   Limitations

This section acknowledges the limitations of the current book in spite of its significant contributions to the body of knowledge. The limitations are as follows:

- First and foremost, the book's scope was restricted solely to the design stage, in terms of the project lifecycle. Therefore, the construction and operation stages were excluded from the investigation though these project phases have considerable impacts on the energy consumption of buildings. This limitation imposed a restriction on the parameters' identification initiative and lifted the occupant behaviour category from the pool of variables brainstormed by the Delphi respondents.

- The technical sophistication of the book led to applying a purposive sampling strategy and inviting only experienced and informed experts on BIM and EED for the Delphi study. Henceforth, pre-screening techniques were conducted due to limitations stemming from the necessity of recruiting experienced and/or informed practitioners in the field. This, in turn, increased the risk of sampling error and selection bias from the specific area. Generally, it is difficult to completely avoid selection bias in research; however, it should be acknowledged and declared by the researchers (17).

- Parametric dataset development indicates the third item of limitation: the data were collected from a parametric energy simulation of a hypothetical low-rise residential building. For this reason, the dataset ranges do not cover mid- and high-rise residential models, and the scope was further restricted to four generic climates of cold, temperate, tropical and hot-arid.
- From the AI side of the study, prediction and classification algorithms were limited to ANN and DT. The optimisation algorithm also focused on GA. Although these types of algorithms were selected based on particular and justified reasons and their different combinations were further investigated, the AI-BIM integration typology of research has yet to be generalisable to other types of AI.
- Revit was the BIM suite deployed in the study and fixed to the framework. This decision was substantiated in view of its powerful parametric engine and popularity within the industry and academic environment. Nevertheless, this choice dictated some degree of restraint on the BIM inclusion of the framework. BIM as a process is driven by an extensive range of software packages, including modelling, analysis, simulation and estimation. Such a wide range of application may require testing exclusively and fitting to the framework.

7.5   Recommendations for Future Studies

The limitations of research pave the way for further studies in the field. Research has a cyclical characteristic and does not end up with an obvious point (18). Therefore, the agenda for future studies could be explained as follows.

In addition to the design stage, the construction and operation phases have undeniable impacts on the sustainability and energy efficiency of the built environment. In order to understand the overall situation of the project lifecycle with respect to the AI-enabled BIM-inherited framework, these phases and their lifecycle assessment (LCA) could be conducted. Studying the design, construction and operation phases on a LCA perspective can provide valuable insights toward their integration with BIM and the data sources required. This notion will significantly affect the number and nature of the variables of building energy and their inclusion in AI and the framework development.

The Delphi study was run through three rounds of a purposive sampling method, mostly from energy and BIM viewpoints. Such empirical study could be more robust with the number of rounds increased to cover the different aspects of research problems, challenges and solutions. Moreover, some intrinsic biases toward the respondents in the non-random sampling can be alleviated by employing random and snowball sampling techniques. To perceive the problem from a broader perspective and in addition to BIM and energy experts, data can be collected from a larger sample with more mixed specialties, such as AI people.

The energy dataset was generated based on the metaheuristic-parametric setting of a hypothetical low-rise residential building, one generic layout design and four climates. The simulation can be extended to mid- and high-rise residential buildings with various design layouts and even to the other types of buildings, such as institutional, office, commercial, industrial and healthcare facilities. More climatic regions should also be evaluated to improve the generalisability of the data. Additionally, real-life building datasets can be utilised via infield data collection or applying instrumental devices to get data from operations. Such considerations in data collection and generation would increase the reliability and validity of the research and its findings.

BIM-inherited EED is recommended for further research and development because of its appreciation in higher dimensions of BIM diffusion into EED. The higher level BIM adopts into EED and sustainable construction in general, the more opportunities will emerge in testing and investigating new areas and approaches toward the sustainable built environment. BIM suites that are capable of this level of adoption should be in the spotlight for verification and comparative studies in order to warrant this new field of research.

References

1 Alwaer H, Clements-Croome DJ. Key Performance Indicators (KPIs) and priority setting in using the multi-attribute approach for assessing sustainable intelligent buildings. Building and Environment. 2010;45(4):799–807.

2 Venkatarama Reddy B, Jagadish K. Embodied energy of common and alternative building materials and technologies. Energy and Buildings. 2003;35(2):129–37.

3 Häkkinen T, Belloni K. Barriers and drivers for sustainable building. Building Research & Information. 2011;39(3):239–55.

4 UNEP. SBCI (Sustainable Building and Climate Initiative). Vision for sustainability on building and construction: United Nation Environment Program; 2014. Available from: www.unep.org/sbci/AboutSBCI/annual_reports.asp.

5 Kibert CJ. Sustainable construction: Green building design and delivery: John Wiley & Sons; 2016.

6 Zanni M-A, Soetanto R, Ruikar K. Defining the sustainable building design process: Methods for BIM execution planning in the UK. International Journal of Energy Sector Management. 2014;8(4):562–87.

7 Azhar S, Carlton WA, Olsen D, Ahmad I. Building information modeling for sustainable design and LEED® rating analysis. Automation in Construction. 2011;20(2):217–24.

8 Kibert CJ. Sustainable construction: Green building design and delivery: Wiley; 2012.

9 Brundtland GH. World commission on environment and development: Our common future: Oxford University Press; 1987.

10 GBCA. The value of green star: A decade of environmental benefits. Green Building Council of Australia, Melbourne, Australia; 2013.

11 Chileshe N, editor. PhD in construction management research: What is original contribution to knowledge? The case of TQM. The 21st Annual ARCOM Conference Association of Researchers in Construction Management, London, UK; September 7–9, 2005.

12 Bourke S. Ph.D. thesis quality: The views of examiners. South African Journal of Higher Education. 2007;21(8):1042–53.

13 Handfield RB, Melnyk SA. The scientific theory-building process: A primer using the case of TQM. Journal of Operations Management. 1998;16(4):321–39.

14 Walker DH. Choosing an appropriate research methodology. Construction Management and Economics. 1997;15(2):149–59.

15 Yi W, Chan AP. Alternative approach for conducting construction management research: Quasi-experimentation. Journal of Management in Engineering. 2013;30(6):05014012.

16 Bartunek JM, Rynes SL. The construction and contributions of 'implications for practice': What's in them and what might they offer? Academy of Management Learning & Education. 2010;9(1):100–17.

17 Ritchie J, Lewis J, Nicholls CM, Ormston R. Qualitative research practice: A guide for social science students and researchers: Sage; 2013.

18 Zou PXW, Sunindijo RY. Research methodology and research: Practice nexus. In: Strategic safety management in construction and engineering. John Wiley & Sons, Ltd; 2015. p. 180–213.

Index

For Product Safety Concerns and Information please contact our EU
representative GPSR@taylorandfrancis.com
Taylor & Francis Verlag GmbH, Kaufingerstraße 24, 80331 München, Germany

9 781032 075549